Bernadette Schreiner

Role of the AAA protease Yme1 in mitochondrial protein quality control

Bernadette Schreiner

Role of the AAA protease Yme1 in mitochondrial protein quality control

A dual role for the AAA protease Yme1

Südwestdeutscher Verlag für Hochschulschriften

Impressum / Imprint
Bibliografische Information der Deutschen Nationalbibliothek: Die Deutsche
Nationalbibliothek verzeichnet diese Publikation in der Deutschen Nationalbibliografie;
detaillierte bibliografische Daten sind im Internet über http://dnb.d-nb.de abrufbar.
Alle in diesem Buch genannten Marken und Produktnamen unterliegen warenzeichen-,
marken- oder patentrechtlichem Schutz bzw. sind Warenzeichen oder eingetragene
Warenzeichen der jeweiligen Inhaber. Die Wiedergabe von Marken, Produktnamen,
Gebrauchsnamen, Handelsnamen, Warenbezeichnungen u.s.w. in diesem Werk berechtigt
auch ohne besondere Kennzeichnung nicht zu der Annahme, dass solche Namen im Sinne
der Warenzeichen- und Markenschutzgesetzgebung als frei zu betrachten wären und
daher von jedermann benutzt werden dürften.

Bibliographic information published by the Deutsche Nationalbibliothek: The Deutsche
Nationalbibliothek lists this publication in the Deutsche Nationalbibliografie; detailed
bibliographic data are available in the Internet at http://dnb.d-nb.de.
Any brand names and product names mentioned in this book are subject to trademark,
brand or patent protection and are trademarks or registered trademarks of their respective
holders. The use of brand names, product names, common names, trade names, product
descriptions etc. even without a particular marking in this works is in no way to be
construed to mean that such names may be regarded as unrestricted in respect of
trademark and brand protection legislation and could thus be used by anyone.

Coverbild / Cover image: www.ingimage.com

Verlag / Publisher:
Südwestdeutscher Verlag für Hochschulschriften
ist ein Imprint der / is a trademark of
AV Akademikerverlag GmbH & Co. KG
Heinrich-Böcking-Str. 6-8, 66121 Saarbrücken, Deutschland / Germany
Email: info@svh-verlag.de

Herstellung: siehe letzte Seite /
Printed at: see last page
ISBN: 978-3-8381-3717-9

Zugl. / Approved by: München, LMU, Diss., 2013

Copyright © 2013 AV Akademikerverlag GmbH & Co. KG
Alle Rechte vorbehalten. / All rights reserved. Saarbrücken 2013

Für meine Lieben

*Steck dir ruhig hohe Ziele und sei voll Zuversicht.
Nur wer zu den Sternen greift, wird seine Wünsche erfüllen.*

TABLE OF CONTENTS

1. **INTRODUCTION** 1
   1.1 Cellular protein quality control systems 1
      1.1.1 Proteins - the worker molecules of the cell...1
      1.1.2 Protein quality control and homeostasis (proteostasis)...........................2
      1.1.3 Failure of protein quality control and homeostasis.................................3
   1.2 Molecular chaperones..3
      1.2.1 The Hsp70 system..4
      1.2.2 The chaperonins 5
      1.2.3 The Hsp90 system..6
      1.2.4 The Hsp100 system..7
   1.3 Proteolytic systems...8
      1.3.1 The AAA protein family..8
      1.3.1 The AAA protease family..9
      1.3.2 LON proteases..10
      1.3.3 ClpP proteases..10
      1.3.4 FtsH proteases..12
   1.4 Mitochondrial biogenesis...12
      1.4.1 Mitochondrial subcompartmentalization 12
      1.4.2 Mitochondrial protein import...13
   1.5 Mitochondrial protein quality control 17
      1.5.1 Protein quality control in the mitochondrial outer membrane 18
      1.5.2 Protein quality control in the mitochondrial matrix...............................18
      1.5.3 Protein quality control in the mitochondrial inner membrane 20
      1.5.4 Protein quality control in the mitochondrial intermembrane space 20
   1.6 Mitochondrial m-and i-AAA protease...21
   1.7 Aim of the present study..26

2. **MATERIALS AND METHODS** 27
   2.1 Molecular biology methods 27
      2.1.1 Strategies for isolation of DNA 27
      2.1.2 Enzymatic editing of DNA 29
      2.1.3 DNA purification and analysis...30
      2.1.4 *E. coli* strains...30
      2.1.5 Plasmids and cloning strategies 32
   2.2 Yeast genetic methods..35
   2.3 Protein biochemistry methods 41
      2.3.1 Analytical methods 41
      2.3.2 Preparation of proteins 44
   2.4 Cell biology methods..46
      2.4.1 NaOH cell disruption 46
      2.4.2 "Rödel's" cell disruption..46
      2.4.3 "Fast Mitoprep"...47
      2.4.4 "Big Mitoprep" 47
      2.4.5 Generation of mitoplasts 48
      2.4.6 Digitonin fractionation of mitochondria 49
      2.4.7 Protease treatment 49
      2.4.8 Aggregation assay 50

2.4.9 Ni-NTA agarose pulldown..50
2.5 Immunological methods **50**
　　2.5.1 Overview of antibodies prepared during this thesis...................................50
　　2.5.2 Further antibodies used in this study..51
　　2.5.3 Generation of specific antisera in rabbits..52
　　2.5.4 Detection of proteins on nitrocellulose membranes by immuno-staining ..53
2.6 Special Methods...**53**
　　2.6.1 Mass spectrometry of elution fractions of Ni-NTA agarose pulldown.......53
　　2.6.2 Identification of aggregating proteins by SILAC and mass spectrometry..54
2.7 Chemicals, consumables and equipment **55**
　　2.7.1 Chemicals..55
　　2.7.2 Consumables 58
　　2.7.3 Equipment 58

3. RESULTS **60**
3.1 Generation and characterization of model substrates for............................**60**
investigating folding in the mitochondrial intermembrane space **60**
　　3.1.1 Generation of an IMS-targeted Cytochrome b_2-DHFR model substrate 60
　　3.1.2 Optimization of induction of model substrate expression 62
　　3.1.3 Expression of model substrate in *S. cerevisiae*..63
　　3.1.4 Verification of the steady state levels of endogenous proteins in cells 64
　　　　expressing the model substrates...64
　　3.1.5 Subcellular localization of model substrates expressed *in vivo* 65
　　3.1.6 Submitochondrial localization of the model substrates 66
3.2 Investigation of the folding behavior of the model substrates **68**
　　3.2.1 *In vivo* protease resistance in the absence and presence of methotrexate ...68
　　3.2.2 Requirements for folding of DHFR in the IMS and matrix........................69
3.3 Identification of potential folding helpers of DHFR in the IMS..................**71**
　　3.3.1 Ni-NTA pulldown and label-free quantification by mass spectrometry.....71
　　3.3.2 Confirmation of Ni-NTA pulldown by western blot and immuno-staining 72
3.4 Generation and characterization of Yme1 deletion strain **74**
　　3.4.1 Growth phenotype of *Δyme1* strain..74
　　3.4.2 Mitochondrial DNA in *Δyme1* strain 75
3.5 Behavior of the model substrate in the absence of Yme1 ....**75**
　　3.5.1 Expression of model substrates in *Δyme1* strain..75
　　3.5.2 Effect of Yme1 on the folding of DHFR 76
3.6 Identification of endogenous Yme1 substrates..**77**
　　3.6.1 Identification of proteins that aggregate in the absence of Yme1 by SILAC
　　　　...77
　　and mass spectrometry 77
　　3.6.2 Endogenous levels of Yme1 substrates in *Δyme1* strain............................85
　　3.6.3 Aggregation of endogenous Yme1 substrates in mitochondria of *Δyme1* ..86
　　　　strain..86
　　3.6.4 Characterization of Mpm1 87
　　3.6.5 Effect of deletion of *YME1* on Mpm1 expression levels 88
　　3.6.6 Aggregation of Mpm1 in the absence of Yme1 89
　　3.6.7 Co-isolation of endogenous substrates of Yme1 with His-tagged Yme1 ...90

4. DISCUSSION **92**
4.1 Folding of the model substrate DHFR in the mitochondrial **92**

    intermembrane space..92
    4.2 Role of Yme1 in folding of the model substrate DHFR................................93
    4.3 Endogenous substrates of the chaperone-like activity of Yme1 95

    4.4 Folding in the mitochondrial intermembrane space - unconventional.......99
    pathways 99
    4.5 The human i-AAA protease 101
    4.6 Protein quality control in the light of neurodegenerative diseases............101
5.   SUMMARY..104
6.   LITERATURE..106
7.   ABBREVIATIONS..123
8.   ACKNOWLEDGEMENTS 123

1. INTRODUCTION

1.1 Cellular protein quality control systems

1.1.1 Proteins - the worker molecules of the cell

In the cell, a multitude of processes are active in parallel at any time, among them transcription and translation, catabolic and anabolic metabolism, and respiration to mention only a few. In all types of cells, proteins are the key factors that execute and maintain these processes. Each protein is synthesized as a linear chain of a defined amino acid sequence. In order to become functionally active, this linear chain has to assume a specific three-dimensional structure, called the native fold. In some cases, an additional level of complexity arises by the assembly of several molecules into a supramolecular quarternary structure. The necessary and sufficient information specifying the final three-dimensional appearance of a protein is contained in its primary sequence (Anfinsen, 1973; Dobson and Karplus, 1999).

Many small proteins can fold spontaneously *in vitro* without the help of accessory factors (Hartl *et al.*, 2011), but the *in vivo* situation is much more challenging. All cellular compartments contain proteins in high concentrations, a situation called molecular crowding (Ellis and Minton, 2006). This creates the danger of nonspecific protein-protein interactions of newly synthesized or misfolded polypeptide chains on their folding pathway (Ellis and Minton, 2006). Unfolded polypeptide chains can accumulate in kinetically trapped conformations. For further folding into a conformation with even lower free energy, trapped conformers have to overcome a free energy barrier. This step is presumably catalyzed by specialized folding helpers, called "molecular chaperones" (Bartlett and Radford, 2009; Hartl *et al.*, 2011). In order to ensure correct protein folding and prevent the deleterious aggregation of unfolded or misfolded proteins, the cell engages an elaborate system of chaperones. In conjunction with proteases that degrade terminally misfolded proteins, chaperones constitute the cellular protein quality control network.

1.1.2 Protein quality control and homeostasis (proteostasis)

For the health and stability of a cell or a multicellular organism, it is crucial that the quality of proteins is maintained at any time. Protein synthesis and degradation also needs to be balanced and adjusted to the prevailing environmental conditions (Chen *et al.*, 2011). Protein homeostasis (proteostasis) is accomplished by the concerted action of protein folding systems and proteolytic systems. They are known as the protein quality control network of the cell (Fig. 1). To a certain extent, these systems can adjust to changing internal or external requirements, thus allowing for a broader survivability of the cell. For instance, heat stress or oxidative stress causes increased unfolding and aggregation of proteins. In response, the cell can switch on transcription and translation of key components of the quality control system. In this way, the capacity of the quality control system is expanded for a short time. This control circuit is named the environmental stress response, after its underlying

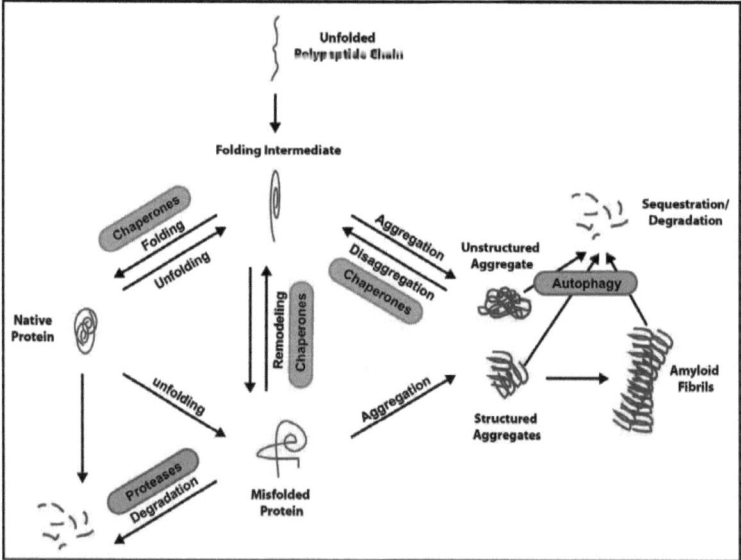

Figure 1. The protein quality control network
From bacteria to eukaryotes, the protein quality control system integrates a network of chaperones (see 1.2) for *de novo* folding of newly synthesized proteins, disaggregation and refolding of misfolded states and a degradation pathways combining specified proteases as the ubiquitin proteasome system and autophagy. Heat shock proteins (Hsps) represent the main class of chaperones whereas AAA proteases constitute the main players in degradation. (Modified from Hartl *et al.*, 2011)

mechanism (Buchberger *et al.*, 2010; Voisine *et al.*, 2010; Chen *et al.*, 2011).

In eukaryotic cells, quality control systems exist in the cytosol and in the organelles. The principle of the environmental stress response applies also to the membrane-enclosed organelles. Elaborate crosstalk must occur between the nucleus and the organelles to generate the organellar unfolded protein response (UPR). An organellar unfolded protein response has been described in the endoplasmic reticulum (Walter and Ron, 2011; Hetz, 2012) and recently also in mitochondria (UPR^{mt}; mitochondrial UPR) (Zhao *et al.*, 2002; Yoneda *et al.*, 2004; Haynes and Ron, 2010). Furthermore, both organelles contain elaborate proteolytic systems that degrade that which can not be (re)folded. In the endoplasmic reticulum (ER), this system is named ERAD (ER-associated degradation) (Romisch, 2005; Bukau *et al.*, 2006).

1.1.3 Failure of protein quality control and homeostasis

The failure of protein quality control systems has severe consequences for the cell. In humans, the dysfunction of critical protein quality control components results in serious disorders, amongst them the neurodegenerative symptom complexes of Morbus Alzheimer, Morbus Parkinson and Chorea Huntington. The neuropathological hallmark of these diseases is the accumulation of toxic protein aggregates in specific cerebral areas. This causes cell death resulting in the characteristic clinical symptoms: dementia and movement disorders (Taylor *et al.*, 2002; Stefani and Dobson, 2003). It is assumed that age-associated decline of the capacity of the protein quality control systems is the reason for the late onset of these typically age-dependent degenerative disorders. One of the major objectives for diagnosis and development of therapeutic treatments is to understand the underlying mechanisms of proteome maintenance (Chiti and Dobson, 2006; Leidhold and Voos, 2007; Balch *et al.*, 2008; Hartl *et al.*, 2011).

1.2 Molecular chaperones

Proteins that help other proteins to reach their native fold but which are not part of the mature three-dimensional structure themselves are called molecular chaperones (Hartl and Hayer-Hartl, 2009). All kingdoms of life contain complex systems of chaperones. Proteins of very different classes belong to the chaperone family. Heat shock proteins (Hsps) are the best known class of chaperones. Hsps were discovered as a class of proteins whose expression is upregulated upon heat shock. However, it subsequently became clear that most heat shock proteins are actually ubiquitously expressed and some are not even upregulated upon cellular stresses (Nollen and Morimoto, 2002). Their functions range from *de novo* protein folding to refolding of denatured proteins, disaggregation of small aggregates, protein trafficking, and targeting of proteins to degradation (Powers *et al.*, 2009; Hartl *et al.*, 2011) (Fig.1). Usually, chaperones use repeated cycles of binding to and release from their substrates. This reaction cycle is regulated by ATP hydrolysis. The classical ATP-dependent chaperones belong to the Hsp70, Hsp60-Hsp10, Hsp90 and, Hsp100 systems.

1.2.1 The Hsp70 system

Members of the Hsp70 family exist in the cytosol of eubacteria (DnaK), in the cytosol of eukaryotes and in all major subcellular compartments of eukaryotic cells: the ER, chloroplasts and mitochondria (Misselwitz *et al.*, 1998; Liu *et al.*, 2001; Mayer and Bukau, 2005). Hsp70 proteins are highly abundant and have a conserved domain structure. The N-terminal nucleotide-binding domain of ~ 44 kDa mediates ATP binding and hydrolysis. The C-terminal peptide binding domain of ~ 27 kDa contains a β-sandwich with the peptide binding cleft and an α-helical segment serving as a lid (Bukau and Horwich, 1998). Unfolded proteins are recognized by means of exposed hydrophobic amino acid side chains and bound in an extended conformation (Erbse *et al.*, 2004; Mayer, 2010).

The substrate affinity of Hsp70 proteins depends on the nucleotide state. In the ATP-bound state the affinity for the substrate is low whereas in the ADP-bound state the affinity is high. The intrinsic ATPase rate of Hsp70s is very low. Co-chaperones of the family of J domain proteins (Hsp40 family) are necessary to stimulate the ATPase activity of the Hsp70s and thus couple substrate binding to ATP hydrolysis (Rowley *et al.*, 1994; Kampinga and Craig, 2010).

ATP hydrolysis by the Hsp70s is accompanied by distinct conformational changes that characterize the transition from the low affinity state to the high affinity state. For release of the substrate, exchange of ADP by ATP is necessary. Binding of

ATP also induces the transition back from the high affinity state to the low affinity state. This step is mediated by nucleotide exchange factors (Bolliger *et al.*, 1994).

1.2.2 The chaperonins

Proteins that fail to assume their native three-dimensional structure with the help of Hsp70 chaperones (Heyrovska *et al.*, 1998) interact with a downstream chaperone system, the chaperonins of the Hsp60 family. Chaperonins are double-ring complexes of ~ 800 kDa forming a central cavity that can accommodate substrates of up to 60 kDa. This cavity shields the substrate protein from the surrounding. The principle of folding within a central cavity has the main advantage that substrate proteins are protected from the aberrant interactions and aggregation. Such incorrect interactions can easily occur due to the molecular crowding in the cell (Brinker *et al.*, 2001).

Chaperonins are subdivided into two classes. The group I chaperonins, called HSP60 in eukaryotes and GroEL in eubacteria, consist of seven subunits per ring and work hand in hand with HSP10 proteins (GroES in eubacteria). The best studied example of group I chaperonins is the bacterial GroEL-GroES system, whose folding cycle has been described in detail (Horwich and Fenton, 2009; Hartl *et al.*, 2011). Hydrophobic stretches of the substrate bind to the apical part of the hydrophobic central cavity of one GroEL ring, the '*cis*' ring. Binding of ATP to the *cis* ring induces conformational changes that enable binding of the GroES lid (Walter, 2002). The wall of the central cavity changes from a hydrophopic state to a hydrophilic state. Thus, the substrate is forced to bury its hydrophobic residues, which promotes folding. Moreover, substrates have a limited range of sterical freedom inside the cavity, and this also favors the formation of more compact and native-like structures (Chakraborty *et al.*, 2010). The substrate remains in the cavity for approximately ten seconds (Xu *et al.*, 1997; Hartl and Hayer-Hartl, 2009). Subsequently, ATP hydrolysis in the *cis* ring and binding of unfolded polypeptide and ATP to the opposite ring (*trans* ring) trigger the release of GroES and the substrate (Horwich *et al.*, 2006). Similar to folding assisted by HSP70, substrates that fail to fold during one ATP cycle of the chaperonin rebind and undergo another round of folding (Horwich *et al.*, 2006; Hartl *et al.*, 2011).

Group II chaperonins (TRiC) exist only in the eukaryotic cytosol and are composed of eight paralog subunits (Frydman, 2001). Group II chaperonins do not cooperate with HSP10 proteins, but have a type of built-in lid that can close the central cavity upon substrate engagement (Douglas et al., 2011; Munoz et al., 2011). A recent study integrated chemical cross-linking, mass spectrometry and combinatorial modeling to determine the subunit arrangement of the TRiC complex (Leitner et al., 2012).

1.2.3 The Hsp90 system

Hsp90 proteins are highly conserved from eubacteria to mammals. Dimerization of two monomers is a prerequisite for functional activity. Each monomer is comprised of an N-terminal nucleotide binding domain, a middle domain that is involved in ATP hydrolysis and substrate binding and a C-terminal dimerization domain. In contrast to the usually unfolded substrates of other chaperone systems, many Hsp90 substrates have a native or near-native state.

ATP binding triggers the dimerization of two Hsp90 protomers. This leads to the formation of a 'molecular clamp' that can bind substrate proteins (Hessling et al., 2009; Hartl et al., 2011). The dimerization also induces hydrolysis of the bound ATP However, the intrinsic ATPase rate of Hsp90 is low and largely modulated by a great number of co-chaperones. The co-chaperones Hop/Sti1 and Cdc37 stabilize the open conformation of Hsp90 thus inhibiting ATP hydrolysis (Roe et al., 2004; Vaughan et al., 2006; Hessling et al., 2009). In contrast, co-chaperone Aha1 binds to the closed conformation and accelerates ATP hydrolysis. P23/Sba1 also binds to the closed conformation, however it inhibits ATP hydrolysis similar to Hop/Sti1 and Cdc37 (Roe et al., 2004). Furthermore, post-translational modifications such as phosphorylation or nitrosylation of amino acid residues in the hinge region between the middle domain and the dimerization domain modulate the ATPase activity of Hsp90 (Retzlaff et al., 2009; Mollapour et al., 2010). The co-chaperones Hop/Sti1, Cdc37 and p23/Sba1 also work as adaptor proteins for substrates of Hsp90.

The substrate binds to Hsp90 in the open conformation. Subsequently, dimerization of the two Hsp90 monomers triggers release of the substrate (Mayer, 2010). The multitude of co-factors enables Hsp90 proteins to handle a large range of substrates and to fine-tune the HSP90 reaction cycle (Scheufler et al., 2000; Hartl et al., 2011).

In eukaryotic cells, HSP90 occupies a branch point in numerous essential signaling pathways, amongst them cell cycle progression, apoptosis and innate immunity (Taipale et al., 2010). Hsp90 proteins have been suggested to promote evolution by buffering the effects of structurally destabilizing mutations, and thus allowing the generation of new variants of proteins (Rutherford and Lindquist, 1998).

The functions of Trap1, the mitochondrial representative of the Hsp90 family, remain largely elusive. They do not seem to overlap with the functions of other Hsp90 chaperones (Dollins et al., 2007; Frey et al., 2007; Leskovar et al., 2008) (Felts et al., 2000). However, it was recently suggested that Trap1 helps to prevent the formation of reactive oxygen species (ROS) and ROS-induced apoptosis (Wandinger et al., 2008).

1.2.4 The Hsp100 system

Hsp100 chaperones belong to the superfamily of AAA+ P-loop ATPases (*A*TPases *a*ssociated with various cellular *a*ctivities). All AAA+ proteins form homo-hexameric ring complexes (Neuwald et al., 1999; Ogura and Wilkinson, 2001; Lupas and Martin, 2002) (see 1.3.1). Each subunit is comprised of one or two highly conserved AAA+ domains. The AAA+ domain is subdivided into a small and a large subdomain. AAA+ domains share the highly conserved Walker A and B motifs that are involved in nucleotide binding (Lupas and Martin, 2002).

ClpB in *E. coli* and its homologs in plants (Hsp101) and lower eukaryotes (Hsp104) mediate disaggregation of protein aggregates by threading them through a central pore. Unfolded substrates are passed on to the Hsp70/Hsp40 system for new folding attempts (Ben-Zvi and Goloubinoff, 2001; Krzewska et al., 2001; Lee et al., 2004; Buchberger et al., 2010; Lotz et al., 2010; Richter et al., 2010). Indeed, Hsp100 chaperones not only deliver disaggregated proteins to the Hsp70/Hsp40 system, but also function in the recovery of damaged Hsp70 proteins after cellular stress (von Janowsky et al., 2005).

Surprisingly, higher organisms lack the Hsp100 system, although an "unfoldase" activity has recently been reported (Bieschke et al., 2009; Murray et al., 2010). It is not yet clear how higher eukaryotes overcome the lack of this system. It has been speculated, however, that the Hsp70 system could have taken over the functions of the Hsp100 system (Ben-Zvi and Goloubinoff, 2001). Indeed, it has recently been shown that the human Hsp70 protein mortalin can mediate disaggregation and thus compensate for the lack of Hsp100 chaperones (Iosefson *et al.*, 2012).

Several representatives of the Hsp100 chaperone family (ClpA, ClpX, HslU in *E. coli*) associate with ring-shaped compartmental peptidases (ClpP, HslV), which will be discussed in detail in 1.3.3.

1.3 Proteolytic systems

1.3.1 The AAA protein family

Proteases involved in protein quality control provide the 'last resort' for terminally damaged proteins that fail to refold and thus constitute a high risk for the cell (Pickart and Cohen, 2004). All thus-far identified proteases that are involved in protein degradation belong to the AAA family of P-loop NTPases. Besides their role in protein degradation, these enzymes occupy hub positions in a multitude of other cellular processes such as mitosis, meiosis, membrane fusion and transport, and DNA replication (Ogura and Wilkinson, 2001).

AAA proteins consist of six or seven subunits forming large ring shaped structures. Each subunit contains one (type I ATPase) or two (type II ATPase) typical AAA domains of approximately 230 amino acids (Kunau *et al.*, 1993; Sauer and Baker, 2011; Langklotz *et al.*, 2012). The highly conserved Walker A and Walker B motifs of the AAA domain are involved in ATP binding and hydrolysis. A conserved lysine residue within the P-loop of the Walker A motif is essential for ATP binding (Neuwald *et al.*, 1999). Mutation of this lysine residue prevents nucleotide binding and thus inactivates the AAA protein (Walker *et al.*, 1982a; Iyer and Aravind, 2004). An aspartate residue within the Walker B motif coordinates a magnesium ion that is essential for ATP hydrolysis (Iyer *et al.*, 2004; Hanson and Whiteheart, 2005).

AAA proteins represent a subfamily of the AAA+ proteins and share some characteristic structural features, distinguishing them from other AAA+ family members. The most prominent of these features is the second region of homology (SRH) (Lupas and Martin, 2002; Frickey and Lupas, 2004) present at the C-terminal part of the AAA domain. Two conserved arginine residues of the second region of homology are involved in ATP hydrolysis and in mediating conformational changes between neighboring subunits (Karata et al., 1999; Ogura et al., 2004).

1.3.1 The AAA protease family

The AAA protease family also belongs to the superfamily of AAA+ proteins. AAA proteases combine proteolytic and ATPase function. Proteolytic and AAA activities can be located on the same or on different polypeptide chains. LON and FtsH proteases contain both activities on one polypeptide chain whereas ClpP proteases associate with separate AAA proteins to form the functional enzyme complex. Each protease family contains specific auxiliary domains that often serve as docking sites for adaptor proteins. In case of FtsH, this "extra" domain anchors the protease to the membrane (Ito and Akiyama, 2005).

The proteolytic sites of AAA proteases are face an "inner chamber" (Tyedmers et al., 2010; Sauer and Baker, 2011; Voos, 2012) and the enzymes are thus designated 'chambered' or 'compartmental' proteases (Bieniossek et al., 2006; Suno et al., 2006). The diameter of the chamber is rather small and thus only unfolded polypeptides are able to enter it. The ATPase activity unfolds the substrates and threads them into the proteolytic chamber.

The unfolding mechanism on the molecular level has been investigated in detail for the bacterial enzyme ClpXP (see 1.3.3). Nucleotide binding to one subunit of the homo-hexamer leads to a rotation between large and small AAA subdomains and this movement is propagated to the neighboring subunit (Lupas and Martin, 2002). ATP hydrolysis drives rigid body movements of the entire AAA ring that are carried forward to the highly conserved central substrate binding loops in the with the GYVG motif. These loops 'roll' from an up position to a down position and thereby translocate the polypeptide chain through the central pore (Hanson and Whiteheart, 2005).

Proteolysis is an irreversible process, and therefore it is strictly controlled. Substrates for degradation are usually marked with specific degradation signals, also designated degradation tags or degrons (Sauer and Baker, 2011). Proteins that are targeted for proteasomal degradation contain bipartite degrons. Such degrons consist of a proteasome binding signal and a degradation initation site.

INTRODUCTION

The major subfamilies of AAA proteases are the LON, Clp and FtsH proteases. They are named after their prokaryotic representatives whose structures, functions and working mechanisms are known in detail.

1.3.2 LON proteases

LON proteases are soluble representatives of the AAA protease family (Schmidt et al., 2009a). They are highly conserved and can be found in the cytosol of eubacteria and in the mitochondrial matrix of eukaryotic cells. LON proteases consist of an N-terminal substrate binding domain, an ATP-binding domain and a C-terminal proteolytic domain (Cha et al., 2010). Therefore, LON proteases contain proteolytic and ATPase activities on a single polypeptide chain. Because they harbor a catalytic lysine-serine diad, LON proteases belong to the class of serine proteases.

LON proteases are essential for the removal of misfolded and oxidatively damaged proteins (Goldberg and Waxman, 1985; Suzuki et al., 1994; Bota et al., 2002; Botos et al., 2004). The bacterial LON protease mediates approximately half of all protein degradation (Ondrovicova et al., 2005; von Janowsky et al., 2005); (Gui and Sauer, 2008). LON substrates carry stretches of hydrophobic amino acid residues that would usually be buried in the native fold of the protein. These residues function as degradation tags for recognition by LON.

Notably, a decrease in the concentration of LON protease in human cells causes defects in the mitochondrial morphology and an increased rate of apoptosis. This emphasizes the fact that an intact proteolytic system is essential for overall survival of the cell (Bota et al., 2002; Bota et al., 2005).

1.3.3 ClpP proteases

ClpP proteases constitute a second class of soluble proteases. They belong to the family of caseinolytic proteases and harbor a characteristic histidine-aspartate-serine catalytic triad. ClpP consists of two homo-heptameric rings where the proteolytic sites are facing an inner tunnel (Kang et al., 2002; Kang et al., 2005; Sauer and Baker, 2011). Distinct from LON proteases, the ATPase activity is not encoded on the same polypeptide chain, and thus ClpP assembles with a homo-hexameric AAA chaperone ring that in turn functions in the recognition of substrates, unfolding, and translocation of the unfolded polypeptide chain into the proteolytic chamber (Kang et al., 2002). Three different partner chaperones are known in bacteria - ClpX, ClpA and ClpC (Baker and Sauer, 2012). ClpA and ClpC contain two AAA domains that seem to function independently as protein unfoldases. However, the biological reason for the presence of two AAA domains is not yet understood (Kress et al., 2009).

Additional factors function as adaptors and target substrate proteins for degradation by binding simultaneously to the substrate and ClpP (Baker and Sauer, 2012). The dimeric adaptor protein SspB targets ssrA-tagged (11-amino acid tail) proteins to ClpXP (Levchenko et al., 2000; Flynn et al., 2004; Lies and Maurizi, 2008). The monomeric adaptor protein ClpS contains a specific binding pocket for 'N-end rule' residues and targets such substrates to ClpAP (Dougan et al., 2002; Erbse et al., 2006; Mogk et al., 2007; Schmidt et al., 2009b).

Further important representatives of the ClpP protease family are the eukaryotic 26S proteasome, the archeal PAN proteasome, and the prokaryotic 20S proteasome (Striebel et al., 2009; Sauer and Baker, 2011). These supermolecular complexes have a modular construction principle that allows for variation in the assembly. The proteolytic core is composed of four stacked rings of ClpP proteases. The proteolytic unit is capped by AAA chaperone rings on both sides. All active proteolytic sites are concentrated in the central cavity, and therefore the substrate can be degraded by multiple proteolytic sites at the same time (Sauer and Baker, 2011).

The substrate recognition process is best investigated for the eukaryotic 26S proteasome. Specific enzymes (E1, E2, E3) transfer and activate polyubiquitin tags on the substrate protein (Elsasser and Finley, 2005). These tags are then recognized by receptor proteins of the proteasome. Only recently, the intrinsic ubiquitin receptor Rpn13 that mediates recognition of polyubiquitinated substrates by the proteasome in cooperation with adaptor proteins was identified (Husnjak et al., 2008). In addition to the polyubiquitin tag, degradation by the 26S proteasome depends on an unstructured initiation site in the substrate protein as a starting point for unfolding and degradation. 26S proteasomes are present in the nucleus and the cytosol, where a subgroup is associated with the ER mediating ER-associated degradation (Smith et al., 2011).

Recently, a similar ubiquitin-like tagging system has been identified in bacteria. Substrates are labeled for degradation by transfer of the small conserved protein Pup, similar to ubiquitinylation in eukaryotes (Pearce *et al.*, 2008). Substrate recruitment and recognition mechanisms of the bacterial 20S proteasome are, however, not yet well understood (Striebel *et al.*, 2009).

1.3.4 FtsH proteases

In contrast to LON and ClpP proteases, the AAA family of FtsH proteases is anchored to the membrane by its auxiliary domain (Karata *et al.*, 2001). The protein family is named after the eubacterial representative FtsH. In cooperation with LON protease, FtsH mediates the degradation of unfolded ssrA-tagged proteins. However, its main task is the protein quality control of membrane proteins (Langklotz *et al.*, 2012).

The interactome of bacterial FtsH comprises HtpX, an ATP-independent endopeptidase that has overlapping substrate specifity with FtsH (Sakoh *et al.*, 2005) and the prohibitin-like proteins HflK and QmcA (Chiba *et al.*, 2006). HflK and QmcA regulate the activity of FtsH. FtsH-like proteases are also found in the membranes of eukaryotic mitochondria or chloroplasts. The two mitochondrial representatives in yeast are located in the inner mitochondrial membrane and will be discussed in detail in 1.6.

1.4 Mitochondrial biogenesis

1.4.1 Mitochondrial subcompartmentalization

Mitochondria are highly complex organelles descending from a bacterial ancestor that was engulfed by the progenitor of a eukaryotic cell (Gray, 1999). For this reason, mitochondria are delineated by two lipid bilayers, the inner and the outer mitochondrial membrane. These membranes define two aqueous compartments, the intermembrane space and the innermost matrix (Palade, 1953). A number of essential cellular processes take place in mitochondria: the tricarboxylic acid cycle (TCA cycle) (Shadel, 2005), heme biosynthesis, iron-sulfur cluster assembly (Lill *et al.*, 2012), β-oxidation of fatty acids, metabolism of certain amino acids and oxidative phosphorylation to mention only a few. Oxidative phosphorylation converts the nutrient-derived energy into the energy carrier molecule ATP that can be utilized by the cell (Dudkina *et al.*, 2010; Nunnari and Suomalainen, 2012). An elaborate set of mitochondrial proteins regulate the morphological dynamics of the organelle (Shaw and Nunnari, 2002). Furthermore, mitochondria also play a key role in apoptosis. They stimulate the intrinsic cell death pathway by release of several pro-apoptotic factors from the mitochondrial intermembrane space (Youle and van der Bliek, 2012). Given the multitude of molecular players involved in these essential processes, an intricate control system is needed to ensure protein quality control and homeostasis.

Remniniscent of their bacterial origin, mitochondria still contain their own genome. The mitochondrial DNA resides in the matrix and encodes only a few proteins (e.g. 8 in *S. cerevisiae*, 13 in *H. sapiens*, and 25 in *A. thaliana*) (Reichert and Neupert, 2004). The vast majority of mitochondrial proteins are, however, encoded in the nucleus and translated on cytosolic ribosomes. Newly synthesized mitochondrial proteins have to be translocated across one or two mitochondrial membranes and sorted to one of the four subcompartments. In order to achieve this task, an elaborate system of protein translocases that recognize, translocate and sort mitochondrial proteins has evolved to guide these transport processes (Mokranjac and Neupert, 2008; Chacinska *et al.*, 2009; Becker *et al.*, 2012). Mitochondrial protein translocases only allow transport of largely, if not completely, unfolded proteins.

1.4.2 Mitochondrial protein import

Nuclear-encoded mitochondrial proteins are translated as precursor proteins harboring specific mitochondrial targeting signals. These targeting signals contain all the information that is necessary and sufficient to target the proteins to the correct submitochondrial compartment. The first known targeting signals are the N-terminal matrix-targeting signals (Neupert and Herrmann, 2007; Mokranjac and Neupert, 2008). The matrix-targeting signals are not conserved in their primary sequence but they all have the potential to form amphipathic helices with one positively charged and one hydrophobic side. The matrix-targeting signal is necessary and sufficient to target the precursor proteins to the mitochondrial matrix where the targeting signal is cleaved off by mitochondrial processing peptidase (Arretz *et al.*, 1991). During the last years, it has become clear that in addition to the N-terminal matrix-targeting signals, a multitude of further targeting signals exist. These involve in particular internal targeting signals that are not cleaved off after import. Most of them are still poorly characterized and it is conceivable that many more will still be found.

INTRODUCTION

The TOM (translocase of the mitochondrial outer membrane) complex forms the major import pore of the outer membrane (Neupert and Herrmann, 2007; Endo and Yamano, 2009) (Fig. 2). The receptor components of the TOM complex recognize mitochondrial precursor proteins by their targeting signals. The TOM complex alone is sufficient for the transport of a subset of outer membrane proteins and some intermembrane space proteins. Transport of all other proteins depends on the concerted action of TOM and additional mitochondrial translocases.

β-barrel proteins of the outer membrane are passed to by the small TIM (translocase of the mitochondrial inner membrane) complexes when they emerge from the TOM complex (Koehler, 2004b). The small TIM complexes escort the β-barrel proteins through the intermembrane space to the TOB/SAM (topogenesis of outer membrane β-barrel proteins/ sorting and assembly machinery) complex in the outer membrane. TOB/SAM mediates the insertion of the β-barrel proteins into the outer membrane (Kozjak *et al.*, 2003; Paschen *et al.*, 2003; Wiedemann *et al.*, 2003).

Only recently, a further system that inserts proteins into the outer membrane was detected. Mim1 and Mim2, two integral outer membrane proteins, mediate the insertion of some single-span, and all known multi-span, outer membrane proteins (Becker *et al.*, 2008). Mim1 and Mim2 interact in the outer membrane to form the MIM (mitochondrial import) complex (Dimmer *et al.*, 2012). MIM mediates the insertion of many components of the TOM complex into the outer membrane and probably also promotes assembly of the TOM complex (Dimmer and Rapaport, 2010).

For import of a subset of small proteins that contain conserved cysteine residues in the form of CX_3C or CX_9C motifs, TOM cooperates with the Mia-Erv disulfide relay system of the intermembrane space. This recently discovered import system introduces disulfide bonds into its substrates (Stojanovski *et al.*, 2008; Deponte and Hell, 2009; Koehler and Tienson, 2009; Herrmann and Riemer, 2012). It is assumed that the introduction of disulfide bonds promotes folding of the entire substrate protein. Immediate folding after translocation through TOM complex also prevents retrograde translocation. Thus, transport through the TOM complex becomes uni-directional.

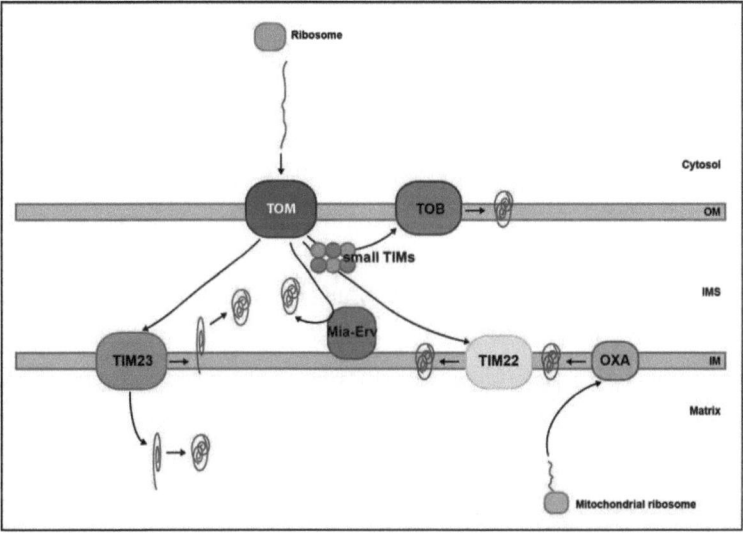

Figure 2. Protein import into mitochondria
Nuclear-encoded proteins with mitochondrial destination are translocated through the outer membrane (OM) by TOM complex. Small TIM complexes escort outer membrane proteins for insertion to the TOB complex in the OM, and members of the carrier family to TIM22 complex in the inner membrane (IM). A subset of IMS proteins is trapped in the intermembrane space (IMS) by introduction of disulfide bonds by the Mia-Erv disulfide relay system. Precursors carrying a matrix-targeting signal (MTS) are passed from the TOM complex to TIM23 in the inner membrane, which mediates translocation across the inner membrane or lateral insertion of precursors with an additional stop-transfer signal into the lipid bilayer. The OXA1 complex of the inner membrane mediates co-translational insertion of proteins encoded in the mitochondrial genome into the inner membrane.

A similar folding trap mechanism was also observed for a group of Mia-Erv-independent intermembrane space proteins. Apo-cytochrome is translocated through the TOM complex and assumes its mature fold upon addition of a heme group in the intermembrane space (Dumont *et al.*, 1991). Sod1 also folds in the intermembrane space upon formation of a disulfide bond by Ccs1 (Field *et al.*, 2003).

Members of the carrier family of inner membrane proteins contain multiple transmembrane domains and are thus very hydrophobic. At the exit of the TOM complex they are received by the small TIM complexes. The small TIM complexes shield the hydrophobic segments of the carrier proteins and guide them to the TIM22 complex of the mitochondrial inner membrane. TIM22 promotes membrane potential-dependent insertion of carrier proteins into the inner membrane (Rehling *et al.*, 2003; Koehler, 2004a).

However, the majoritiy of mitochondrial precursors follow the main import pathway through the TOM complex in the outer membrane and the TIM23 complex in the inner membrane. The TIM23 translocase can work in two functional modes, for insertion of precursor proteins into the inner membrane on the one hand, and for translocation of precursor proteins into the matrix on the other hand (Mokranjac and Neupert, 2010; Marom *et al.*, 2011). Most precursor proteins of this pathway carry N-terminal matrix-targeting signals that are cleaved off by the mitochondrial processing peptidase in the matrix. Membrane potential is needed for translocation of the matrix-targeting signal across the inner membrane. For translocation of the mature part of precursor proteins, the membrane potential is not sufficient. The ATP-dependent import motor of TIM23 provides the driving force for import of the mature part of the precursor protein into the matrix (Neupert and Brunner, 2002). Precursors that are targeted to the inner membrane contain hydrophobic stop-transfer signals downstream of the matrix-targeting signal. The stop-transfer signal induces lateral opening of the TIM23 complex and release into the lipid bilayer of the inner membrane. These proteins can then either stay anchored in the inner membrane by their transmembrane domain or undergo a second cleavage at the level of the inner membrane. Proteins that are cleaved twice are released as soluble proteins into the intermembrane space.

Some inner membrane proteins use an alternative sorting pathway to reach the inner membrane. They enter the matrix through the TIM23 complex and are then sorted back into the inner membrane by the OXA1 complex of the inner membrane. This pathway is referred to as the conservative sorting pathway as it is apparently inherited from gram-negative bacteria, the endosymbiontic ancestors of mitochondria. Recent evidence suggests that the lateral insertion and conservative sorting pathways can cooperate for insertion of multitopic inner membrane proteins (Bohnert *et al.*, 2010).

The OXA1 complex handles yet another set of proteins, namely those that are encoded in the mitochondrial DNA (Herrmann and Neupert, 2003; Mokranjac and

Neupert, 2009). Proteins that are encoded in the mitochondrial genome are translocated in a co-translational manner. It is assumed that the main driving force for membrane insertion is provided by the mitochondrial ribosome that pushes the nascent polypeptide chain into the OXA1 translocase (Herrmann and Neupert, 2000).

1.5 Mitochondrial protein quality control

Almost all mitochondrial proteins have to cross one or two membranes in order to reach their destination. Translocases transport only largely unfolded proteins and therefore folding can only occur at the compartment of destination. For this purpose, intrinsic mitochondrial folding systems have evolved. The inner mitochondrial membrane and the matrix are also not accessible to the cytosolic protein quality control systems. Thus, not only intrinsic folding systems but also intrinsic quality control systems are needed in the diverse mitochondrial subcompartments (Baker and Haynes, 2011). Intact mitochondria are crucial for the integrity of the whole cell, and mitochondrial dysfunction correlates with aging and the onset of a multitude of neurodegenerative diseases (Rugarli and Langer, 2006; Hartl *et al.*, 2011). This emphasizes that a functional mitochondrial protein quality control system is essential for cell survival.

Elaborate and complex systems of mitochondrial chaperones and proteases represent the first level of mitochondrial quality control. The second level of mitochondrial quality control is based on mitochondrial dynamics (Hoppins *et al.*, 2007; Otera and Mihara, 2011). Mitochondrial fusion enables exchange of intact components between mitochondria, whereas mitochondrial fission can separate severely damaged mitochondrial areas. Such areas are subsequently removed by mitophagy, thus preventing a toxic effect of terminally damaged mitochondria on still intact neighboring mitochondria (Tanaka *et al.*, 2010; Wang and Klionsky, 2011; Youle and Narendra, 2011). The largest effects are conveyed by apoptosis, the third level of quality control. In the case of extensive cellular damage that overwhelms the first two levels of protein quality control, mitochondria can trigger the intrinsic apoptosis pathway by release of pro-apoptotic factors from the mitochondrial intermembrane space (Wasilewski and Scorrano, 2009; Martin, 2010). The following paragraphs will focus on the first level of mitochondrial protein quality control and discuss the principles of the most important systems.

1.5.1 Protein quality control in the mitochondrial outer membrane

Proteins of the outer mitochondrial membrane are degraded by cytosolic proteasomes in a process named outer mitochondrial membrane associated degradation (OMMAD) (Neutzner et al., 2007) similar to ER-associated degradation (ERAD) (Smith et al., 2011). We have only recently begun to identify the molecular factors that are involved in OMMAD has started. In mammalian cells, the E3 ubiquitin ligase Parkin transfers a poly-ubiquitin tag to outer membrane proteins such as mitofusins that are involved in mitochondrial fusion. A cytosolic AAA protease (Cdc48/p97/Vcp), also involved in ER-associated degradation, extracts ubiquitinated outer membrane proteins and transfers them to cytosolic proteasomes.

1.5.2 Protein quality control in the mitochondrial matrix

A set of "classical" chaperones and proteases ensures protein quality control in the mitochondrial matrix (Fig. 3). They represent the homologs of their cytosolic counterparts. Mitochondrial Hsp70, Ssc1, is involved in the import of newly incoming polypeptides and subsequently assists in their folding (Matouschek et al., 2000; Neupert and Herrmann, 2007). The distinction between the two functional modes of Ssc1 is mediated by Hsp40/DnaJ co-chaperones (Horst et al., 1997). Tim14-Tim16 complex (D'Silva et al., 2003; Mokranjac et al., 2007) is the main regulator of the intrinsic ATPase activity of Ssc1 within the TIM23 complex, whereas Mdj1 regulates the ATPase activity of Ssc1 during the folding of proteins in the matrix. Mge1 serves as a nucleotide exchange factor in both functional modes (Voos and Rottgers, 2002).

Yeast mitochondria express three types of Hsp70s. Ssc1 is the most abundant, assisting translocation and folding processes, Ssq1 is involved in the assembly of iron-sulfur clusters (Schilke et al., 1999) and the third representative is the largely uncharacterized Ssc3 (Baumann et al., 2000). Notably, mutations in mortalin, the human mitochondrial Hsp70, have recently been suggested to play an important role in the generation of diverse diseases such as cancer and age-related neuropathies (Yaguchi et al., 2007).

Polypeptide chains that fail to assume their native three-dimensional structures with the help of Ssc1 are passed on to the Hsp60-Hsp10 system (Fig. 3) (see 1.2.2).

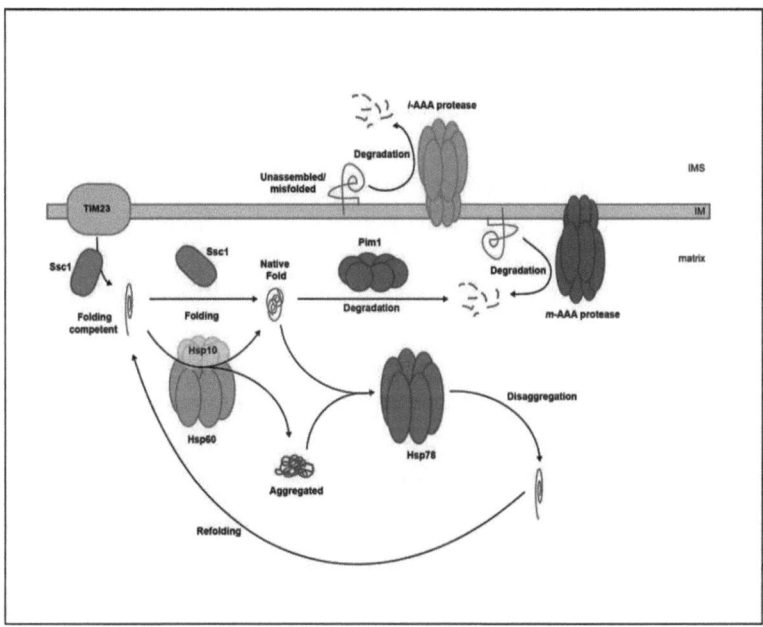

Figure 3. Protein quality control systems of mitochondrial matrix and IMS
Mitochondrial Hsp70, Ssc1, is involved in the import and subsequent folding of newly incoming precursor proteins. Downstream of Ssc1, the Hsp60-Hsp10 complex provides a shielded inner cavity for folding. Misfolded proteins are extracted from aggregates by the Hsp78 unfoldase and passed on to Ssc1 or Hsp60-Hsp10 for renewed folding. Terminally damaged proteins are degraded by the soluble AAA+ protease Pim1. Degradation of inner membrane proteins is accomplished by the concerted action of two membrane anchored AAA+ proteases: *i*- and *m*-AAA protease.

Substrate proteins can undergo folding inside the interior cavity of Hsp60 (Hartl, 1995; Mayer, 2010). Proteins that fail to fold in this process or get unfolded or damaged at a later time point of their life have a high tendency to aggregate. Such misfolded proteins and aggregates represent a high risk for mitochondrial functionality.

A matrix-located protease system can remove such terminally damaged proteins. Pim1, a member of the LON family of AAA proteases (see 1.3.2), degrades misfolded or otherwise damaged soluble matrix proteins (Bender *et al.*, 2011; Wagner *et al.*, 1994) (Fig. 3). Pim1 receives its substrates from Hsp70 and Hsp60 chaperone systems. Hsp70 and Hsp60 promote degradation of folding-resistant substrates by

Pim1. It is, however, still unclear how the chaperones 'decide' whether a unfolded substrate undergoes another round of folding or is targeted for degradation (Chen *et al.*, 2011). Hsp78, a member of the ClpB/Hsp100 family (see 1.2.4), functions as a disaggregase. It contains a AAA+ domain, thus belonging to the AAA+ superfamily. In contrast to the Clp family of AAA ATPases, Hsp78 unfolds proteins and passes them on to Hsp70 or Hsp60 for refolding or to proteases for degradation (Schmitt *et al.*, 1995; Rottgers *et al.*, 2002).

1.5.3 Protein quality control in the mitochondrial inner membrane

Two membrane-anchored proteases of the AAA family, the *i*-AAA and *m*-AAA protease (see 1.3.1) are involved in degradation of proteins at the levels of the inner membrane (Fig. 3). The catalytic domains of *i*-AAA and *m*-AAA protease face the intermembrane space and the matrix, respectively (Koppen and Langer, 2007; Tatsuta and Langer, 2009). Both were shown to be involved in the degradation of non-assembled proteins that are peripherally associated to, or integrated into, the inner mitochondrial membrane (Arlt *et al.*, 1996; Leonhard *et al.*, 1996; Korbel *et al.*, 2004). *i*-AAA and *m*-AAA protease have overlapping substrate specificity, and it was suggested that substrate discrimination occurs only on the basis of the opposing orientation of the two proteases in the inner membrane.

1.5.4 Protein quality control in the mitochondrial intermembrane space

In the mitochondrial intermembrane space, no members of the classical chaperone families are known. Obviously, atypical folding helpers could exist, but for most intermembrane space proteins it is not clear how they assume their mature fold. In comparison to the volume of the mitochondrial matrix, the intermembrane space constitutes a relatively small compartment. However, the integrity of the proteome of the intermembrane space is highly important. Such essential processes as the exchange of diverse ions and metabolites and oxidative phosphorylation depend on an intact proteome of the intermembrane space. Furthermore, in higher eukaryotes, the release of the intermembrane space protein cytochrome c triggers the caspase cascade inducing apoptosis. Therefore, in order to maintain proteostasis in the intermembrane space, appropriate quality control mechanisms must exist in this compartment.

However, the key factors and mechanisms that are involved still remain largely elusive (Herrmann and Hell, 2005; Neupert and Herrmann, 2007).

Only recently, the above-mentioned Mia-Erv disulfide relay system was discovered and shown to mediate vectorial movement of proteins containing conserved cysteine residues across the outer membrane. For some of the rather small Mia-Erv substrates, it has been shown that introduction of disulfide bonds by Mia40 is sufficient to trigger folding of the entire polypeptide chain (Hell, 2008; Koehler and Tienson, 2009; Sideris and Tokatlidis, 2010; Herrmann and Riemer, 2012).

A similar folding trap mechanism has been observed for another group of intermembrane space proteins. Cytochrome c, for instance, is imported through the TOM complex as an apo-protein. It assumes its mature fold upon addition of a heme group by cytochrome c heme lyase (Dumont et al., 1991). Similarly, Sod1 assumes its mature fold upon introduction of a disulfide bond and Cu^{2+} by Ccs1 (Field et al., 2003). Thus, for a subset of intermembrane space proteins, addition of a cofactor or introduction of disulfide bonds seems to be sufficient to trigger folding of the entire protein. For all other intermembrane space proteins, amongst them very large ones with a complex structure, the folding process remains unknown. Neither folding helpers nor energy sources are known.

The i-AAA protease Yme1 of the mitochondrial inner membrane belongs to the AAA protease family (see 1.3.1). Yme1 exposes its catalytic domain into the intermembrane space (Koppen and Langer, 2007; Baker et al., 2011) and was shown to mediate ATP-dependent degradation of intermembrane space and inner membrane proteins (Leonhard et al., 1996; Leonhard et al., 2000). Several endogenous substrates of the proteolytic activity of Yme1 were identified recently (Weber et al., 1996; Leonhard et al., 2000; Kominsky et al., 2002; Dunn et al., 2006; Graef et al., 2007; Potting et al., 2010; Elliott et al., 2011). Already thirteen years ago, Yme1 was suggested to exert a chaperone-like activity. Its isolated AAA domain has been shown to refold denatured model substrates *in vitro* (Leonhard et al., 1999). However, *in vivo* evidence for a chaperone-like activity of Yme1 is so far lacking.

1.6 Mitochondrial m-and i-AAA protease

m- and i-AAA protease, the two AAA proteases of the inner mitochondrial membrane, share the conserved domain structure of AAA proteases (Fig. 4). The N-terminal regions harbor one (*i*-AAA) or two (*m*-AAA) transmembrane segments. The AAA domain contains two conserved motifs, the nucleotide-binding Walker A motif (GXXXXGKS/T) and the magnesium-binding Walker B motif ($X_hX_hX_hX_hDEXX$) (Walker et al., 1982b; Lupas et al., 1997). Furthermore, the AAA domain contains the so-called second region of homology (SRH) (Beyer, 1997; Patel and Latterich, 1998; Ogura and Wilkinson, 2001). SRH is essential for ATP hydrolysis and mediates the contact between two neighboring subunits in the hexamer (Karata et al., 1999; Karata et al., 2001; Ogura et al., 2004). The AAA domain is followed by a proteolytic domain of the M41 thermolysin metalloprotease family. The proteolytic domain contains a conserved zinc binding HEXGH consensus motif (Rawlings and Barrett, 1995).

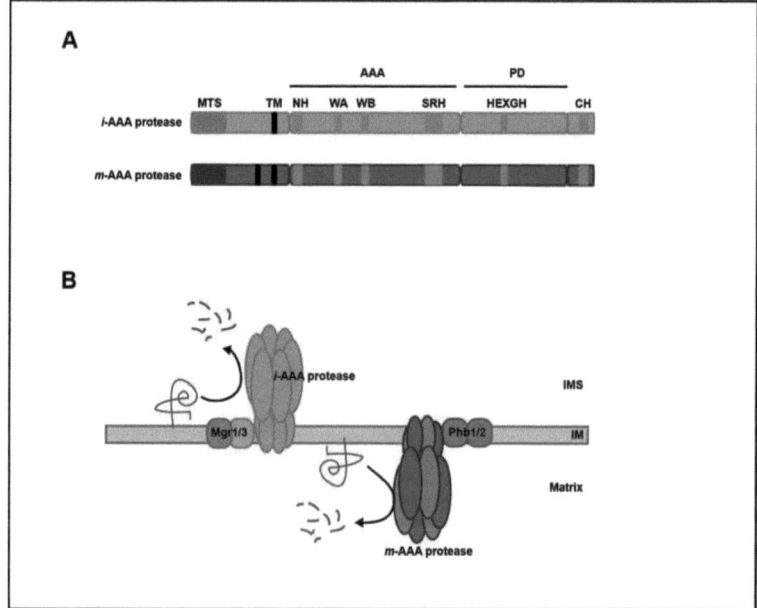

Figure 4. Mitochondrial *i*- and *m*-AAA+ protease
A) *i*- and *m*-AAA protease show the typical domain structure of the family of AAA+ proteases, with an N-terminal AAA+ motif including Walker A and B motifs (WA, WB) and the second region of homology (SRH), and a C-terminal proteolytic domain (PD) of the thermolysin family carrying the conserved proteolytic HEXGH motif. **B)** The catalytic domains of *i*- and *m*-AAA proteases face the IMS and the matrix, respectively. The potential adaptor proteins Mgr1 and Mgr3 are found in a supercomplex with the *i*-AAA protease, whereas *m*-AAA protease forms a supercomplex with prohibitins 1 and 2 (Phb1/2). MTS, matrix-targeting signal; TM, transmembrane domain; NH, N-terminal helices; CH, C-terminal helices. Modified from Koppen and Langer, 2007.

In yeast, the deletion of i- or m-AAA protease induces an elevated rate of mitochondrial DNA escape to the nucleus (Thorsness and Fox, 1993). The deletion strains are respiratory deficient although this defect is not due to the mitochondrial DNA escape. These strains show a significant growth defect on non-fermentable medium at high temperatures and on fermentable medium at low temperatures (Campbell et al., 1994; Weber et al., 1996). Moreover, the deletions strains are 'petite negative' and therefore cannot grow without mitochondrial DNA, in contrast to wild-type *S. cerevisiae* strains (Thorsness and Fox, 1993; Campbell et al., 1994). Early electron microscope studies revealed that the deletion strains have altered mitochondrial architecture. The typical mitochondrial network is destroyed, mitochondria are grossly swollen and located in close proximity to vacuoles (Campbell et al., 1994; Campbell and Thorsness, 1998). This might be due to an increased turnover of mitochondria in the deletion strains, although evidence for this notion is lacking. The double deletion of both AAA proteases is lethal, thus emphasizing the overlapping substrate specificity of the two proteases.

The yeast m-AAA protease consists of two alternating subunits, Yta10 and Yta12. Three Yta10 and three Yta12 subunits form a hetero-hexameric ring structure of approximately 1000 kDa (Arlt et al., 1996) (Fig. 4). Yta10 and Yta12 are homologs and share 53 % sequence identity. The m-AAA protease is helps mediate several regulatory steps during mitochondrial biogenesis (Esser et al., 2002; Tatsuta et al., 2007). In addition to the turnover of misfolded and damaged proteins, m-AAA protease functions as a processing peptidase. It cleaves the mitochondrial ribosome protein MrpL32, which is a crucial step for integration of MrpL32 into the mitochondrial ribosome (Nolden et al., 2005). Furthermore, m-AAA protease functions as a membrane dislocase in the maturation process of Ccp1. The m-AAA-dependent dislocation of Ccp1 from the membrane is the prerequisite for processing of Ccp1 by rhomboid protease Pcp1 (Esser et al., 2002; Tatsuta et al., 2007). Recently, m-AAA protease was found in a supercomplex with prohibitins, protein and lipid scaffolds of the inner membrane (Steglich et al., 1999). Two homologous membrane-anchored proteins, Phb1 and Phb2, have been identified in all eukaryotic genomes (Berger and Yaffe, 1998). Based on their deletion phenotypes, it is likely that Phb1 and Phb2 exert a regulatory function during protein turnover by m-AAA protease (Steglich et al., 1999; Nijtmans et al., 2000; Langer et al., 2001).

INTRODUCTION

The importance of an intact *m*-AAA protease is underlined by the finding that mutations in the genes encoding paraplegin or Afg3l2, the two subunits of mouse *m*-AAA protease, cause two severe neurological diseases: hereditary spastic paraplegia and spinocerebellar ataxia (Rugarli and Langer, 2006; Martinelli *et al.*, 2009; Tatsuta and Langer, 2009).

The mitochondrial *i*-AAA protease consists of six Yme1 subunits that form a homo-hexameric ring structure similar to the *m*-AAA protease. Yme1 is anchored to the inner membrane by one transmembrane domain in the N-terminal segment and exposes its catalytic domains to the intermembrane space (Koppen and Langer, 2007; Gerdes *et al.*, 2012) (Fig. 4). Distinct from the *m*-AAA protease, no dislocation or processing activities of Yme1 have been reported so far. However, several endogenous substrates of the proteolytic activity of Yme1 have been identified during the last few years.

Cox2 and Nde1 are the first proteolytic substrates of Yme1 that were confirmed by *in vivo* studies (Nakai *et al.*, 1995; Pearce and Sherman, 1995; Weber *et al.*, 1995; Lemaire *et al.*, 2000; Kominsky *et al.*, 2002; Augustin *et al.*, 2005; Graef *et al.*, 2007) Nde1 is a soluble protein residing in the mitochondrial intermembrane space, providing NADH for the respiratory chain (Luttik *et al.*, 1998). Cox2 is encoded in the mitochondrial genome (Cooper *et al.*, 1991), and is the terminal member of the electron transport chain in the inner mitochondrial membrane. Hence, the localization of both verified Yme1 substrates is compatible with an the localization of the catalytic sites of Yme1. Recently, Ups1 and Ups2, two intermembrane space proteins that are involved in lipid metabolism, were shown to be degraded by Yme1 (Osman *et al.*, 2009a; Tamura *et al.*, 2009; Potting *et al.*, 2010). Similarly, prohibitins, lipid scaffolds of the inner membrane, were identified as proteolytic substrates of Yme1 (Potting *et al.*, 2010). However, further endogeneous substrates of Yme1 remain to be identified. Mgr1 and Mgr3, two integral inner membrane proteins, form a supercomplex of approximately 1000 kDa with Yme1. The deletion of either protein is not lethal and even the double deletion strain is viable. However, the proteolytic activity of Yme1 seems to be impaired in the deletion strains, suggesting an adaptor-like function of Mgr1 and Mgr3 (Dunn *et al.*, 2006; Dunn *et al.*, 2008)

Recently, a role of Yme1 in import into the mitochondrial intermembrane space was postulated. The import of ectopically expressed human polynucleotide phosphorylase into the intermembrane space of isolated yeast mitochondria was dependent on Yme1. In the absence of Yme1, non-imported polynucleotide phosphorylase accumulated in the cytosol (Rainey et al., 2006). Upon import, the polynucleotide phosphorylase remained stable in the intermembrane space and was not degraded by Yme1. This observation suggests that Yme1 may have a function apart from its proteolytic activity

As mentioned before, already thirteen years ago, *in vitro* experiments raised the idea of a chaperone-like function of Yme1. The recombinantly expressed and purified AAA domain of Yme1 was shown to mediate refolding of two model substrates, rhodanese and DHFR (Leonhard *et al.*, 1999). *In vivo* experiments proving this hypothesis are, however, missing so far.

1.7 Aim of the present study

The aim of this study was to gain insight into the folding of proteins in the mitochondrial intermembrane space. For this purpose, mouse dihydrofolate reductase, a protein that is frequently used to analyze translocation into mitochondria and subsequent folding processes, was to be expressed in the intermembrane space of mitochondria in intact yeast cells. The folding of dihydrofolate reductase was to be investigated under different conditions. Furthermore, its folding helpers would be determined and, finally, endogenous substrates of these folding helpers would be identified and analyzed.

2. MATERIALS AND METHODS

2.1 Molecular biology methods

2.1.1 Strategies for isolation of DNA

2.1.1.1 Isolation of genomic DNA from *S. cerevisiae*

For isolation of genomic DNA, an overnight culture of the respective yeast strain was incubated at 30 °C while shaking (130 rpm) (Rose et al, 1990). Cells were pelleted by centrifugation (2500 x g, 5 min, RT), washed with 20 ml H_2O and disrupted by resuspension in 1ml breaking buffer (100 µg/ml zymolyase 20T, 1M sorbitol, 100 mM EDTA). After incubation for 1 h at 37 °C, the cells were washed with 1ml 1 M sorbitol and 100 mM EDTA, pelleted by centrifugation, resuspended and then incubated in 1 ml lysis buffer (50 mM Tris/HCl, 20 mM EDTA, 1 % (w/v) SDS, pH 7.5) for 30 min at 65 °C. The cell solution was supplemented with 400 µl 5 M potassium acetate and incubated for 1 h on ice. DNA was separated from the protein fraction by centrifugation (20000 x g, 15 min, 4 °C). The supernatant was transferred to a new tube, DNA precipitated with isopropanol and pelleted by centrifugation (37000 x g, 10 min, 2 °C). After washing with 70 % ethanol, the DNA pellet was dried for 10 min at RT and resuspended in 100 µl sterile H_2O. After determination of concentration with a Nanodrop 2000c spectrophotometer (Promega), the DNA was stored at -20 °C.

2.1.1.2 Isolation of plasmid DNA from *E. coli*

Plasmid DNA was isolated in small scale or large scale from *E. coli* based on the principle of alkaline lysis (Birnboim and Doly, 1979) using the "PureYield" Plasmid Midiprep System (Promega). *E. coli* clones containing the appropriate plasmid were inoculated in 50 ml LB^{Amp} medium and grown overnight at 37 °C with shaking (130 rpm, Infor cell shakers). Bacteria were harvested by centrifugation for 5 min at 5000 rpm in JA-20 rotor (Beckman centrifuge) and resuspended in 6 ml "Cell

Resuspension Solution" (10 mM EDTA (pH 8), 50 mM Tris/HCl (pH 7.5), 100 µg/ml RNase A). 6 ml "Cell Lysis Solution" (0.2 M NaOH, 1 % (w/v) SDS) and in a second step 10 ml "Neutralization Solution" (4.09 M guanidine hydrochloride (pH 4.2), 759 mM potassium acetate, 2.12 M glacial acetic acid) were added. The samples were mixed gently after each step and centrifuged for 10 min at 10000 rpm (JA-20 rotor). The supernatant of the centrifugation step was loaded onto a "clearing column" on top of a "binding column" and passed through the column by vacuum. The "clearing column" was removed and the "binding column" was washed with 5 ml "Endotoxin Removal Wash" followed by 20 ml "Column Wash Solution" (8.3 mM Tris/HCl (pH 7.5), 0.04 mM EDTA (pH 8), 60 mM potassium acetate, 60 % ethanol). After the liquid was passed through completely, the vacuum was continued for 30 s to dry the membrane. DNA was eluted from the column with 600 µl ddH$_2$O, concentration was determined with a Nanodrop 2000c spectrophotometer and DNA was stored at -20 °C.

2.1.1.3 Amplification of DNA by polymerase chain reaction

Specific DNA fragments were amplified in a thermo cycler (eppendorf) by polymerase chain reaction (PCR) (Saiki et al., 1985; Saiki et al., 1988). One reaction cycle includes the following steps: denaturation of template strands, annealing of single strand oligonucleotides (primers; obtained from Metabion, Martinsried/ Germany), elongation of primers by thermo stable DNA-polymerase. A typical 50 µl reaction mix contained 5 µl supplied 10x PCR buffer, 4 µl dNTP mix (dATP/ dCTP/ dGTP/ dTTP; 10 mM), 6.25 µl 20 µM forward and reverse primer, ca. 100 ng gDNA or 10 ng plasmid DNA as template and 1 U *Taq* polymerase. When the PCR product was to be used for subsequent cloning, the proof-reading polymerase *Pfu* was used instead. For further analysis, PCR products were separated by agarose gel electrophoresis and isolated by gel extraction (see 2.1.3.1, 2.1.3.2). For diagnostic PCR of clones, 1 µl of single *E. coli* colonies resuspended in 15 µl sterile H$_2$O were used as template.

PCR cycler protocol

Step	Cycles	Temperature, Time
Initial denaturation	1	95 °C, 3 min
Denaturation	30	95 °C, 30 s
Annealing of primers	30	ca. 55 °C, 30 s
Elongation of primers	30	72 °C, 1 min per 1kb
Final elongation	1	72 °C, 10 min
Cooling	1	4 °C, for ever

2.1.2 Enzymatic editing of DNA

2.1.2.1 Restriction digestion of DNA

Sequence-specific digestion of DNA was achieved by using highly specific restriction endonucleases in the supplied buffer (New Englang Biolabs). 2 - 5 U of enzyme were applied per 1 µg DNA. For an analytical approach, DNA samples were digested for 30 min in 10 µl reaction volume. For a preparative approach, DNA samples were digested for 2.5 h in 50 µl reaction volume. Incubation temperature was always 37 °C. Digested DNA fragments were purified by agarose gel electrophoresis and subsequent gel extraction (see 2.1.3.1, 2.1.3.2).

2.1.2.2 Ligation of DNA fragments

The concentrations of digested vector and insert were estimated on an agarose gel. Linearized vector (100 - 200 ng) and 5 - 10 molar excess of DNA insert were mixed in a 20 µl reaction containing 2 µl T4 DNA ligase and corresponding 10x T4 ligation buffer (50 mM Tris/HCl, 10 mM $MgCl_2$, 1 mM DTT, 1 mM ATP, 5 % (w/v) PEG-8000, pH 7.6) (both New England Biolabs). Ligation reactions were incubated for 3 h at room temperature or at 16 °C overnight. 0.5 - 1 µl of ligation reaction were transformed into electro-competent *E. coli* cells (see 2.1.4.4).

2.1.3 DNA purification and analysis

2.1.3.1 Agarose gel electrophoresis

For separation according to size, DNA samples were subjected to horizontal agarose gels electrophoresis. 0.8 % (w/v) agarose gels, supplemented with 0.5 g/ml ethidium bromide, were run in TAE buffer (40 mM Tris/ acetate, pH 7.4, 20 mM natrium acetate, 1 mM EDTA), and DNA samples were mixed with 5x concentrated loading dye (50 % (w/v) glycerol, 0.25 % (w/v) bromphenol blue, 0.25 % (w/v) xylencyanol). The gels were run at 10-15 V/cm and DNA fragments were visualized by UV light.

2.1.3.2 Extraction of DNA from agarose gels

A piece of agarose gel containing the DNA fragment of interest was cut out with a scalpel under UV light (366 nm) and put into 1.5 ml reaction tube. DNA was isolated from the gel using the Promega gel extraction kit through silica columns according to the manufacturer's instructions.

2.1.4 *E. coli* strains

2.1.4.1 Overview of *E. coli* strains

Strain	Genotype	Reference
MH1	*K12 derivative, melBLid, ΔnhaA1, kan+, ΔnhaB1, cam+, ΔlacZY, thr1, MH1*	Harel-Bronstein et al., 1995
BL21 (DE3)	*B F– dcm ompT hsdS(r_B^- m_B^-) gal*	Studier and Moffatt, 1986
XL1-Blue	*supE44, hsdR17, recA1, endA1, gyrA96, thi-1, relA1, lac⁻, F'[proAB⁺, lacI^q lacZΔM15, Tn10(tet^r)]*	Stratagene

For cloning, I used the *E. coli* strain MH1, which is a derivative of MC1061 strain and harbors streptomycin resistance. For recombinant protein expression, *E. coli* strains BL21 (DE3) or XL1-blue were used. The BL21 (DE3) strain has a chromosomal copy of T7 RNA polymerase controlled by *lacUV5* promoter. The strain lacks the proteases Lon and OmpT that stabilizes the overexpressed proteins. XL1-blue cells express the *lac* repressor. Addition of IPTG leads to its inactivation and thus to the induction of downstream proteins.

2.1.4.2 Cultivation of *E. coli*

E. coli cultures were grown in LB medium in the presence of a suitable antibiotic at 37 °C and 130 rpm (Infor shakers). Pre-cultures were inoculated from a culture plate or directly from the corresponding glycerol stock. For preparation of stocks, *E.coli* cell suspension was mixed 1:1 with 87 % glycerol, snap-frozen in liquid nitrogen and stored at -80 °C.

LB (Luria-Bertani)/ LB^{Amp}/ LB^{Kan}	10 g/l bacto-trypton, 5 g/l bacto-yeast extract, 5 g/l NaCl; optional: 100 µg/ml ampicillin/ kanamycin

For preparation of the according agar plates, 2 % (w/v) agar was added before autoclaving and plates were poured when the medium was cooled down to ca. 60 °C.

2.1.4.4 Preparation of electro-competent *E. coli* cells

1 ml of 50 ml overnight E. coli culture was inoculated into 500 ml LB medium and incubated at 37°C in a shaker until an OD_{600} of ca. 0.5 was reached. The culture was incubated on ice for 30 min followed by sedimentation through centrifugation for 5 min at 5000 rpm (JA-10 rotor, Beckman Coulter) at 4 °C. The cell pellet was washed with 500, 250 and 50 ml of ice-cold, 10 % (v/v) glycerol solution. The pellet was resuspended in 500 µl 10 % glycerol, and 40 µl aliquots were stored at -80 °C.

2.1.4.5 Transformation of *E. coli* with plasmid DNA by electroporation

One µl of ligation reaction or of 1:100 diluted plasmid DNA was added to a 40 µl solution of electro-competent cells (see 2.1.4.4). The mixture was transferred to an ice-cold 2 mm electroporation cuvette, put into "GenePulser" (Bio-Rad) electroporator and transformed by a singular short voltage pulse at 2.5 kV/ 400 Ω/ 25

MATERIAL AND METHODS

µF. Afterwards, the cells were immediately diluted in 1 ml pre-warmed LM medium, transferred to a 1.5 ml reaction tube and incubated for 30 min at 30 °C shaking at 130 rpm. Cells were pelleted, resuspended in 100 µl of residual supernatant and plated on LB plates supplemented with the required antibiotics. The plates were incubated over night at 37 °C and ca. 15 single colonies analyzed by diagnostic PCR.

2.1.5 Plasmids and cloning strategies

2.1.5.1 Overview of plasmids used for expression in *E. coli*

Plasmid	Reference
pET28a+-Yme1_AAA-His$_6$	This thesis
pMAL-cRI-Imp1	This thesis

2.1.5.2 Cloning strategies

a) Cloning of Yme1_AAA-His6 into pET28a+

The coding sequence for amino acid residues 250 - 525 of Yme1 was amplified from genomic DNA by PCR using Yme1_AAA_for and Yme1_AAA_rev primers and cloned into the pET28a+ bacterial expression vector (Novagen) upstream of the plasmid-encoded His$_6$-tag using *Nco*I and *Xho*I restriction sites.

Yme1_AAA_for	5'-CC CCC ATG GGG GGT TTT AAA TAC ATC ACA G-3'
Yme1_AAA_rev	5'-CCC CTC GAG CTT TCT CTC AGC ACC C-3'

b) Cloning of Imp1 into pMAL-cRI

The coding sequence of Imp1 was amplified from genomic DNA by PCR using Imp1_Mal_for and Imp1_Mal_rev primers and cloned into the pMAL-cRI bacterial expression vector (NEB) downstream of the plasmid-encoded MBP-tag using *Pst*I and *Hind*III restriction sites.

Imp1_Mal_for	5'-CCC CTG CAG ACT GAG ACG AGG GGA GAA TC-3'
Imp1_Mal_rev	5'-CCC CCC AAG CTT TCA GTT GCT CTT AGC CTG CAC ATC TAG-3'

2.1.5.3 Overview of plasmids used for transformation into *S. cerevisiae*

Yeast strain	Reference
pYES2-Cyb2(1-107)DHFR^{WT} (= IMS-DHFR^{WT})	Popov-Celeketic *et al.*, 2011
pYES2-Cyb2(1-107; Δ19)DHFR^{WT} (= matrix-DHFR^{WT})	This thesis
pYES2-Cyb2(1-107)DHFR$^{mut(7/42/49)}$ (= IMS-DHFR^{mut})	This thesis
pYES2-Cyb2(1-107; Δ19)DHFR$^{mut(7/42/49)}$ (= matrix-DHFR^{mut})	This thesis
pYES2-Cyb2(1-107)DHFR^{WT}His$_9$ (= IMS-DHFR^{WT}His)	Popov-Celecétic et al., 2011
pYES2-Cyb2(1-107; Δ19)DHFR^{WT}His$_9$ (= matrix-DHFR^{WT}His)	This thesis
pRS314-Yme1	This thesis
pRS314-His$_9$-Yme1	This thesis

2.1.5.4 Cloning strategies

a) Cloning of Cyb$_2$(1-107)-DHFR^{mut}(-His$_9$) and Cyb$_2$(1-107; Δ19)-DHFR^{mut}(-His$_9$) into pYES2

The coding sequence of DHFR^{mut} (*Mus musculus*) containing point mutations at position 7 (cysteine to serine), 42 (serine to cysteine) and 49 (asparagine to cysteine) was amplified by PCR from the plasmid pGEM4Su9-DHFR^{mut} using forward primer BamDHFRC7S_for and HindDHFR_rev or HindDHFR_His$_9$ reverse primers, respectively. DHFR^{WT} inserts were excised from plasmids pYES2-Cyb$_2$(1-107)DHFR^{WT} and pYES2-Cyb2(1-107; Δ19)DHFR^{WT} by restriction digest with

BamHI and EcoRI and replaced with the PCR-amplified inserts DHFR^{mut} and DHFR^{mut}His9 using BamHI and EcoRI restriction sites in the primers.

BamDHFRC7S_for	5'-CCC GGA TCC ATG GTT CGA CCA TTG AAC TCC-3'
DHFR_Eco_rev	5'-TTT GAA TTC TTA GTC TTT CTT CTC GTA GAC TTC-3'
DHFRHis$_9$_Eco_rev	5'-TTT GAA TTC TTA ATG GTG ATG GTG ATG GTG ATG GTG ATG GTC TTT CTT CTC GTA GAC TTC-3'

b) Cloning of N-terminal His-tagged Yme1 into pRS314

The plasmid pRS314-Yme1 was amplified by PCR using Yme1_N-His_for and Yme1-N-His_AgeI_rev primers, which anneal after base triplet 50 of the YME1 sequence coding for a serine. The His$_9$-tag was encoded by the two primers and thus introduced behind the presequence of Yme1. Successful cloning was determined by restriction digest of the amplified PCR product with AgeI, whose restriction site was introduced by the reverse primer.

Yme1_N-His_for	5'-CAT CAC CAT CAC GAA AAG AAT AGC GGT GAA ATG CCT CCT AAG-3'
Yme1-N-His_AgeI_rev	5'-GTG ATG GTG ATG GTG AGA ATA AAA CCG GTA GAA CTT CTT TGA TC-3'

c) Amplification of mitochondrial DNA in the Δyme1 strain

Towards addressing the DNA content of mitochondria in *Δyme1* strain, Atp8, encoded in the mitochondrial genome, was amplified with Atp8_for and Atp8_rev primers by PCR. The PCR reaction was separated on a 0.8 % agarose gel and DNA bands were visualized under UV-light.

Atp8_for	5'-TAT ATT ACA TCA CCA TTA GAT C-3'
Atp8_rev	5'-AGC CCA GAC ATA TCC CTG AAT G-3'

All clonings were finally confirmed by sequencing (Metabion, Martinsried).

2.2 Yeast genetic methods

2.2.1 Overview of *S. cerevisiae* strains used

a) Wild-type strains

Yeast strain	Genotype	Reference
YPH499	Mat a ura3-52 lys2-801_amber ade2-101_ochre trp1-Δ63 his3-Δ200 leu2-Δ1	(Sikorski and Hieter, 1989)
W303-1A	w303 Mat α ade2-1; his3-11; leu2, 112; trp1; ura23-53; can1-100	(Sherman and Wakem, 1991)

b) Strains generated by homologous recombination

Yeast strain	Reference
YPH499ΔYME1::KANMX4	This thesis
YPH499Yme1-His$_6$	This thesis
YPH499ΔMGR1::HIS3	This thesis
YPH499ΔMGR3::HIS3	This thesis
YPH499Mpm1-myc	This thesis

2.2.2 Homologous recombination in *S. cerevisiae*

a) Deletion of YME1 gene

YME1 was deleted by homologous recombination with the corresponding PCR product, amplified from pFA6KANMX4 (Wach *et al.*, 1997) using Yme1delta_for and Yme1delta_rev primers in the haploid yeast strain YPH499. To select for positive clones, transformed yeast cells were grown on medium containing kanamycin. Homologous recombination was confirmed by PCR, and the absence of Yme1 by "Fast Mitoprep".

Yme1delta_for	5'-TAA TTA TAA TAC ATT GTG GAT AGA ACG AAA ACA GAG ACG TGA TAG CGT ACG CTG CAG GTC GAC-3'
Yme1delta_rev	5'-GTC TTG AGG TAG GTT CCT TCA TAC GTT TAA CTT CTT AGA ATA AAA ATC GAT GAA TTC GAG CTC-3'

MATERIAL AND METHODS

c) Chromosmal His₇-tagging of C-terminus of Yme1

Yme1_His_for primer, consisting of the last 45 3' bases of *YME1* and 18 bases of pYM5 plasmid, and Yme1_His_rev primer, consisting of the 45 first bases of YME1 3'-UTR followed by 18 further bases of pYM10 plasmid, were used to amplify a PCR fragment from pYM10 containing 5' to 3' the following sequences: the last 45 3' bases of *YME1*, His₇-coding sequence, HIS cassette and 45 bases of the beginning of the 3'-UTR. The PCR fragment was transformed into the YPH499 wild-type strain, and homologous recombination confirmed by PCR. Transformants were selected on medium lacking histidine.

Yme1_His_for	5'-GAT ATA GGC GAT GAT AAA CCC AAA ATT CCT ACA ATG TTA AAT GCA CAC CAT CAC CAT CAC CAT CAC-3'
Yme1_His_rev	5'-GGT GTT ATG AAG CAA AAG CGA AAC CGA CCA GAA AAG AAC AAA GCA TTC ATC GAT GAA TTC GAG CTC G-3'

b) Deletions of MGR1 and MGR3 genes

MGR1 and *MGR3* genes were deleted by homologous recombination in the haploid yeast strain YPH499 with the corresponding PCR products amplified by Mgr1delta_for and Mgr1delta_rev or Mgr3delta_for and Mgr3delta_rev primers. PCR products contained the auxotrophic *HIS3* marker cassette and short sequences homologous to the flanking regions of *MGR1* or *MGR3* loci. Selection for positive clones was performed on medium lacking histidine. Homologous recombination was confirmed by PCR with Mgr1con_for and Mgr1con_rev primers or Mgr3con_for and Mgr3con_rev primers.

Mgr1delta_for	5'-CAT CCT CCT CCA TTC CCT CTC CTT TTC CAA TTA CCG TAA TAA AAG CGT ACG CTG CAG GTC GAC-3'
Mgr1delta_rev	5'-AAG GAT TTA ATA TAC GCA CGG TAC AAC TAA GCA ATC CGC AAA GAC ATC GAT GAA TTC GAG CTC-3'

Mgr3delta_for	5'-CAG GAA GAT CTC AGT TTA ACA GGC TAA AAG TCC CTC CTT TTC GGT CGT ACG CTG CAG GTC GAC-3'
Mgr3delta_rev	5'-TTG AAA TAT TTA TTA TTT TTG TCT TCC TTT ATT TCC TTT ATT GTG ATC GAT GAA TTC GAG CTC
Mgr1con_for	5'-TTC AAA CAC AAT AGT TGT TCG -3'
Mgr1con_rev	5'-TCT CCA AAG GGC AAA GAA ACC-3'
Mgr3con_for	5'-AAC TGT TCA TTG CTT TGT TCC-3'
Mgr3con_rev	5'-TTT TTT CAA ATT GGG TTT AGG-3'

c) Chromosmal myc-tagging of the C-terminus of Mpm1

Mpm1_myc_for primer, consisting of the last 45 3' bases of *MPM1* and 18 bases of pYM5 plasmid, and Mpm1_3'-UTR_rev primer, consisting of the 45 first bases of MPM1 3'-UTR followed by 18 further bases of pYM5 plasmid, were used to amplify a PCR fragment from pYM5 containing 5' to 3' the following sequences: last 45 3' bases of MPM1, myc-coding sequence, HIS cassette and 45 bases of the beginning of 3'-UTR. The PCR fragment was transformed into the YPH499 wild-type strain, and homologous recombination confirmed by PCR. Transformants were selected on medium lacking histidine.

Mpm1_myc_for	5'-AGT CCC CAG GTG AAG CAT AAA GTG GTG AGT GTT GAC GAA GAC ATT CGT ACG CTG CAG GTC GAC-3'
Mpm1_3'-UTR_rev	5'-CAT ATT GTG TAA GAT ATG AGT AAA AAA AGG AAA CGA AAA TAT GTC ATC GAT GAA TTC GAG CTC-3'

2.2.3 Strains generated by transformation with yeast expression vectors

Yeast strain	Reference
YPH499 + pYES2-Cyb2(1-107)DHFR^{WT} (= IMS-DHFR^{WT})	This thesis
YPH499 + pYES2-Cyb2(1-107; Δ19)DHFR^{WT}	This thesis

(= matrix-DHFR^{WT})	
YPH499 + pYES2-Cyb2(1-107)DHFR$^{mut(7/42/49)}$ (= IMS-DHFR$^{mut(7/42/49)}$)	This thesis
YPH499 + pYES2-Cyb2(1-107; Δ19)DHFR$^{mut(7/42/49)}$ (= matrix-DHFR$^{mut7/42/49}$)	This thesis
YPH499 + pYES2-Cyb2(1-107)DHFR^{WT}His (= IMS-DHFR^{WT}His)	This thesis
YPH499 + pYES2-Cyb2(1-107; Δ19)DHFR^{WT}His (= matrix-DHFR^{WT}His)	This thesis
YPH499ΔYME1::KANMX4 + pYES2-Cyb2(1-107)DHFR^{WT} (= IMS-DHFR^{WT})	This thesis
YPH499ΔYME1::KANMX4 + pYES2-Cyb2(1-107; Δ19)DHFR^{WT} (= matrix-DHFR^{WT})	This thesis
YPH499ΔYME1::KANMX4 His$_6$ Yme1 + pYES2-Cyb2(1-107)DHFR^{WT} (= IMS-DHFR^{WT})	This thesis
YPH499ΔYME1::KANMX4 + His$_6$-Yme1 + pYES2-Cyb2(1-107; Δ19)DHFR^{WT} (= matrix-DHFR^{WT})	This thesis
YPH499Yme1-His$_6$ + pYES2-Cyb2(1-107)DHFR^{WT} (= IMS-DHFR^{WT})	This thesis
YPH499Yme1-His$_6$ + pYES2-Cyb2(1-107; Δ19)DHFR^{WT} (= matrix-DHFR^{WT})	This thesis
YPH499ΔMGR1::KANMX4	This thesis
YPH499ΔMGR3::KANMX4	This thesis
YPH499ΔMGR1::KANMX4	This thesis

+ pYES2-Cyb2(1-107)DHFR^{WT} (= IMS-DHFR^{WT})	
YPH499ΔMGR3::KANMX4 + pYES2-Cyb2(1-107)DHFR^{WT} (= IMS-DHFR^{WT})	This thesis
YPH499ΔMGR1::KANMX4 + pYES2-Cyb2(1-107; Δ19)DHFR^{WT} (= matrix-DHFR^{WT})	This thesis
YPH499ΔMGR3::KANMX4 + pYES2-Cyb2(1-107; Δ19)DHFR^{WT} (= matrix-DHFR^{WT})	This thesis

2.2.4 Cultivation of *S. cerevisiae*

Yeast strains were grown according to standard methods (Sherman, 1991). For stocks, yeast cells were taken from a plate with an inoculation loop, transferred into 15 % (v/v) glycerol, snap-frozen in liquid nitrogen and stored at -80 °C. For cultivation, yeast strains from the stock solution were spread on YPD or selective agar plates. Selective plates lacked one or more particular selection markers so that only successfully transformed cells containing the plasmid would grow. After incubation at 30 °C, liquid cultures were inoculated from these plates into 50 ml of medium and incubated at 30 °C at 130 rpm shaking (Infor cell shakers). The cell culture was grown in the logarithmic phase for 2-3 days before inoculating the large cultures. For activation of the *GAL* promoter of yeast expression plasmid pYES2, 0.5 % galactose (w/v) was added two hours before harvesting.

2.2.5 Media for cultivation of *S. cerevisiae*

Lactate medium	3 g yeast extract, 1 g KH$_2$PO$_4$, 1 g NH$_4$Cl, 0.5 g CaCl$_2$ x 2 H$_2$O, 0.5 g NaCl, 1.1 g MgSO$_4$ x 6 H$_2$O, 0.3 ml 1 % FeCl$_3$, 22 ml 90 % lactic acid, H$_2$O ad 1 l, pH 5.5 adjusted with KOH; supplemented with 0.1 % glucose
YPD medium	10 g yeast extract, 20 g bacto-peptone, H$_2$O to 930 ml, pH 5.5 adjusted with KOH; supplemented with 2 % glucose

MATERIAL AND METHODS

YPG medium	10 g yeast extract, 20 g bacto-peptone, H$_2$O to 930 ml, pH 5.5 adjusted with KOH; supplemented with 3 % glycerol
YPGal medium	10 g yeast extract, 20 g bacto-peptone, H$_2$O ad 930 ml, pH 5.5 adjusted with KOH; supplemented with 2 % galactose
SD medium	1.74 g yeast nitrogen base, 5 g (NH$_4$)$_2$SO$_4$, 20 g glucose, H$_2$O ad 1 l
SLac medium	1.7 g yeast nitrogen base, 5 g (NH$_4$)$_2$SO$_4$, 22 ml 90 % lactic acid, H2O ad 1 l, pH 5.5 adjusted with KOH
SILAC medium	SLac medium containing heavy Lysine (L-Lysine-^{13}C$_6$, ^{15}N$_2$ hydrochloride)

Amino acids for selective media were prepared separately. 10 mg/ml solutions of histidine, leucine, lysine and 2 mg/ml solutions of uracil and adenine were autoclaved for 20 min at 120 °C. 10 mg/ml solutions of tryptophan were not autoclaved but filtered for sterilization. For preparation of the corresponding plates, 2 % (w/v) agar was added before autoclaving and plates were poured when the medium was cooled down to ca. 60 °C.

2.2.6 Lithium acetate transformation of *S. cerevisiae*

A pre-culture of the appropriate yeast strain was inoculated, grown overnight and diluted to an OD$_{600}$ of 0.1 - 0.2 in 50 ml the following morning. Cells were grown for two cell divisions to an OD$_{600}$ of 0.5 - 0.6 and pelleted by centrifugation in sterile 50 ml tubes for 5 min at 3000 x g. The cell pellet was washed with sterile water, resuspended in 1 ml 100 mM lithium acetate and transferred to an 1.5 ml reaction tube. After centrifugation for 15 sec at 13200 rpm (eppendorf table-top centrifuge 5415D) the cell pellet was resuspended in ca. 400 µl 100 mM lithium acetate to a final volume of 500 µl. 50 µl cell suspension per transformation was prepared in separate 1.5 ml reaction tubes and centrifuged for 5 min at 3000 x g. After removing the lithium acetate supernantant, 240 µl 50% polyethylene glycol, 36 µl 1M lithium acetate, 5 µl 10 mg/ml single strand DNA, 0.1 - 10 µg plasimd or PCR product and 60 µl H$_2$O were added one after the other and then mixed thoroughly on the vortexer for 1 min. After incubation for 30 min at 30 °C and 200 rpm, followed by 25 min incubation at 42 °C, cells were sedimented for 15 sec at 8000 rpm and the transformation supernatant was removed. Cells were resuspended in 100 µl sterile H_2O streaked out on appropriate selective plates. In the case of kanamycin resistance, cells were incubated for 2 - 3 hours at 30 °C before plating.

2.2.7 Test for growth phenotype of *S. cerevisiae*

The yeast strains of interest were inoculated into 20 ml pre-culture and diluted to an OD_{600} of 0.1 - 0.2 the next morning. After growth for two cell divisions to an OD_{600} of 0.5 - 0.6, 0.1 OD_{600} was collected, washed in sterile H_2O and resuspended in 100 µl sterile H_2O. Starting from this cell suspension, four serial 1:10 dilutions were prepared and 2 µl of each dilution was spotted on YPD and YPG plates. Four copies of each plate were prepared for incubation at 16 °C, 24 °C, 30 °C and 37 °C. Growth was monitored every day and pictures of the plates were taken as soon as the cells of highest dilution started growing.

2.3 Protein biochemistry methods

2.3.1 Analytical methods

2.3.1.1 SDS-polyacrylamide gel electrophoresis (SDS-PAGE)

Proteins were separated on discontinuous SDS-PAGE according to their size (Laemmli et al., 1970). For large gels, the running gel was 9 x 15 x 0.1 cm and the stacking gel 1 x 15 x 0.1 cm in size. For mini gels the Bio-Rad Mini-Protean II apparatus (7 x 7.2 x 0.075 cm; stacking gel: 1 x 7.2 x 0.075 cm) was used. The concentration of acrylamide-bisacrylamide (Serva) in the running gel was chosen appropriate to the sizes of to be separated proteins and was usually between 8 and 14 %. Before pouring the gel, the glass plates were sealed with 2 % (w/v) Agar-Agar (Roth) in running buffer. Protein samples to be analyzed were dissolved in Laemmli buffer (120 mM Tris/HCl pH 6.8, 4 % SDS, 20 % glycerol, 0.01 % bromophenol blue with or without 5 % (v/v) β-mercaptoethanol and boiled for 3 min at 95 °C before loading on the gel. Large gels were run at 35 mA (EP601 power supply/ GE

Healthcare) and mini gels at 25 mA until the bromophenol blue front reached the bottom of the running gel.

2.3.1.2 Coomassie Brilliant Blue staining of SDS gels

Proteins that were separated by electrophoresis were briefly washed in H_2O before staining by incubation of the gel in Coomassi Brilliant Blue (40 % methanol, 10 % acetic acid, 0.1 % Coomassi Brilliant Blue R-250) for ca. 60 min. Consequently, the gel was destained by repeated incubation in 40 % methanol, 10 % acetic acid and then washed in water again.

2.3.1.3 Drying of Coomassie-stained gels

The CBB-stained gel was assembled between two plastic sheets of in water-soaked gel drying films (Promega), fixed by two rectangular plastic frames and dried over night in a fume hood.

2.3.1.4 Transfer of proteins onto nitrocellulose membranes (western blot)

Proteins separated by electrophoresis were transferred onto nitrocellulose according to semi-dry method (Khyse-Anderson, 1984). Two blotting papers (neoLab), one nitrocellulose membrane (Whatman) and the gel were soaked in blotting buffer (20 mM Tris, 150 mM glycine, 0.02 % SDS, 20 % methanol) and assembled to a sandwich in the following order: blotting paper, nitrocellulose membrane, gel, blotting paper. The sandwich was assembled between two graphite electrodes (made in-house by the departmental workshop) and the negatively charged proteins were transferred onto the nitrocellulose membrane by applying constant current (250 mA) for 1 hour. A wet blot sandwich consisted of blotting paper, nitrocellulose membrane and the gel was fixed between two grids transfer took place in a chamber filled with blotting buffer (Towbin et al., 1979). Transfer was performed for 1.5 hours at 400 mA and 4 °C. Transferred proteins on the nitrocellulose membrane were stained with Ponceau S. Immuno-staining of proteins on the membranes and detection of ECL signals was performed as described in 2.5.6.

2.3.1.5 Protein precipitation by trichloracetic acid (TCA) method

Proteins were precipitated from a solution by addition of 72 % (w/v) TCA to a final concentration of 12 % TCA in the sample. Samples were mixed vigorously and put at -20 °C for 30 min prior to high speed centrifugation (20000 rpm/ 12154-H rotor/ Sigma) at 4 °C for 20 min. After removing the TCA supernatant and washing the protein pellet with ice-cold acetone, the pellet was dried for 10 min at room temperature, dissolved in 2x Laemmli buffer with 5 % (v/v) β-mercaptoethanol and heated at 95 °C for 5 min.

2.3.1.6 Determination of protein concentration

For determination of protein concentration, duplicates of 3 and 6 µl of protein solution were supplemented with 1 ml of 1:5 diluted Bradford Reagent (Bio-Rad). After incubation at room temperature for 10 min, the absorbance at 595 was measured and protein concentration determined by a calibration curve derived from absorbances of protein solution of known concentrations (0; 1,5; 3; 6; 12; 24 µg/ml) of bovine IgG standard.

2.3.1.7 Autoradiography

For visualization of radioactively labeled proteins on nitrocellulose membrane, the membranes were dried under red light and exposed to X-ray films (Kodak Bio Max MM). After an appropriate time of exposure, protected from light by special Kodak film cassettes, the films were developed in a Gevamatic 60 developing machine (AGFA-Gevart). The exposure time was adapted to the signal intensities, and ranged from one day to several weeks.

2.3.1.8 Detection and quantification of ECL and radioactive signals on films

Signals on sensitive films were detected and scanned with an ImageScanner (GE Healthcare) and quantified using the accompanying Master 1D Elite software (GE Healthcare).

MATERIAL AND METHODS

2.3.2 Preparation of proteins

2.3.2.1 Overexpression of recombinantly expressed Yme1-AAA-His$_6$ and MBP-Imp1

The AAA domain (amino acid residues 250 - 525) of *S. cerevisiae* Yme1 coupled to C-terminal His6-tag was expressed in the BL21 (DE3) *E. coli* strain from the pET28a+ expression vector (Novagen). Similarly, N-terminally MBP-tagged Imp1 was as well expressed in the BL21 (DE3) *E. coli* strain from the pMAL-cRI vector (New England Biolabs). Overnight cultures of transformed BL21 (DE3) were used to inoculate 4 l main cultures and grown further at 37 °C to an OD$_{600}$ 0.8. Expression of the recombinant protein was induced by addition of IPTG to a final concentration of 0.5 mM and grown for an additional 2.5 h at 37 °C. 1 ml samples of cell culture before induction and after 2.5 h induction were taken out, spun down and resuspended in 100 µl 2x Laemmli with 5 % (v/v) β-mercaptoethanol per 1 OD$_{600}$. Samples were loaded on mini gels and stained with Coomassie to monitor the induction of protein expression. The residual cell culture was harvested by centrifugation at 5000 rpm (JA-10 rotor, Beckman centrifuge) and 4 °C for 5 min, washed with H$_2$O and transferred to 50 ml reaction tubes. Cells were pelleted by a further centrifugation step for 10 min at 4 °C and 4000 rpm (SX4250 rotor/ Beckman X-22R Benchtop Falcon Centrifuge) and either immediately processed further or stored at -20 °C.

2.3.2.2 Purification of recombinantly expressed His-tagged AAA domain of Yme1

The pellet of *E. coli* cells overexpressing *S. cerevisiae* Yme1 AAA-His6 was resuspended in buffer (150 mM KCl, 50 mM Tris, 15 mM imidazol, pH 8, 1 mM PMSF). Lysozyme was added to 1 mg/ml final concentration and the suspension was incubated at 4 °C for 30 min. Cells were disrupted by sonication (Branson Sonifyer 250) with 12 x 12 sec pulses (duty cycle 80 %, setting 4) on ice and 18 sec breaks on ice in between. Cell suspension was transferred to JA20 centrifugation tubes and unopened cells were pelleted by centrifugation at 15000 rpm (JA 20 rotor, Beckmann) and 4 °C for 15 min. The supernatant was loaded onto 2 ml Ni-NTA column. After washing with three column volumes of buffer (150 mM KCl, 50 mM Tris, 15 mM imidazol, pH 8, 1 mM PMSF), the His-tagged protein was eluted with 250 mM imidazol in the same buffer and collected in 15 separate 1 ml fractions. The protein content of each fraction was tested by Ponceau S staining of 2 µl drops of the fractions on a nitrocellulose membrane. The exact protein concentration in the peak elution fractions 4 - 7 was determined by Bradford assay. Samples of 30 μg of recombinant protein per lane were loaded onto 14 % SDS gels. After electrophoresis, proteins were transferred onto nitrocellulose membrane by western blot and visualized by Ponceau S staining. Protein bands containing the overexpressed His-tagged AAA domain of Yme1 were cut out, destained and dissolved in DMSO prior to injection into rabbit once per month (see 2.5.3).

2.3.2.3 Coupling of peptides to SulfoLink beads

The peptide that was used for injection into the rabbit for antibody production was also used for affinity purification of the specific antibodies according to ThermoScientific protocol. For this purpose, 2 ml of SulfoLink Resin were transferred with a wide-bore pipette to an empty column, resulting in a final volume of 1 ml of settled beads. The column was equilibrated with 4 column volumes of coupling buffer (50 mM Tris, 5 mM EDTA-Na, pH 8.5). The lyophilized peptide was dissolved in coupling buffer (2 ml) and added to the column. The peptide was given time for binding to the SulfoLink resin by over-end-mixing at room temperature for 15 min and incubation at room temperature without mixing for a further 30 min. The column was then washed with 3 column volumes coupling buffer and the non-specific binding sites were blocked by mixing with one bed-resin 50 mM L-Cysteine-HCl in coupling buffer for 15 min and further incubation for 30 min, both at room temperature. The column was washed with six column volumes 1 M NaCl and two column volumes of storage buffer (0.1 M Tris, 0.5 M NaCl, pH 8) before being stored in storage buffer with 0.05 % natrium azide at 4 °C.

2.3.2.4 Purification of antibodies

For purification of specific antibodies from the serum, the prepared SulfoLink column was equilibrated to room temperature and in case the column was used for the first time, it was washed with 10 ml 0.1 M glycine, pH 2.5 and 10 ml 0.1 M Tris, pH 8.8. After equilibration with 10 ml 10 mM Tris, pH 7.5, 6 ml of serum diluted with 24 ml 10 mM Tris, pH 7.5, 1 mM PMSF were loaded onto the column. The column was washed with 10 ml 10 mM Tris, pH 7.5 and 10 ml 10 mM Tris, pH 7.5, 0,5 M NaCl

MATERIAL AND METHODS

before elution in three steps was performed. First, antibodies were eluted with 10 ml 0.1 M citrate pH 4.0 and collected in 1 ml fractions that were equilibrated with 200 µl 1 M Tris, pH 8.8. Collected 1 ml fractions of the second elution step with 10 ml 0.1 M glycine, pH 2.5 were equilibrated with 150 µl 1 M Tris, pH 8.8 and the collected 1 ml fractions of the third elution step with 10 ml 0.1 M natrium phosphate, pH 11 were equilibrated in 150 µl 1 M glycine, pH 2.2. Nitrocellulose membranes with 25 µg of alternately loaded wild-type and $\Delta yme1$ mitochondria were immuno-stained with samples of all elution fraction diluted 1:250 in milk. Usually, most proteins were eluted with glycine and acidic pH. After use, SulfoLink columns were washed with 10 ml 0.1 M Tris pH 8.8 and stored in storage buffer with 0.05 % natrium azide for further purifications.

2.4 Cell biology methods

2.4.1 NaOH cell disruption (Kushnirov, 2000)

50 ml pre-culture was grown over night, diluted to an OD_{600} 0.2 in the morning and grown for 2 cell divisions until an OD_{600} of 0.6. 2 OD_{600} of culture were harvested by centrifugation at 14000 rpm (table top Eppendorf centrifuge) and room temperature for 5 min. The cell pellet was resuspended in 200 µl 0.1 M NaOH and incubated for 5 min at room temperature. Disrupted cells were pelleted by centrifugation at 14000 rpm (12154-H rotor/ Sigma) at room temperature for 5 min, resuspended in 50 µl 2x Laemmli with 5 % (v/v) β-mercaptoethanol per 1 OD_{600} and 0.5 - 1 OD_{600} were loaded per lane onto appropriate SDS-PA gels.

2.4.2 "Rödel's" cell disruption (Horvath and Riezman, 1994)

A second method for disruption of yeast cell is 'Rödel's' cell lysis. Yeast cells were grown and harvested as described for NaOH lysis. The cell pellet was resuspended in 50 µl 'Rödel's' mix (1.85 M NaOH, 7.4 % β-mercaptoethanol, 20 mM PMSF), vigorously mixed on the vortexer (VF2/ Janke und Kunkel) and incubated for 15 min on ice. Proteins were precipitated with TCA as described under 2.3.1.5 and analysed by SDS-PAGE, western blot and immuno-staining.

2.4.3 "Fast Mitoprep"

Yeast strains were grown overnight, diluted in the morning and grown to OD_{600} of 0.6 - 0.8. and 5 - 10 OD_{600} were collected in 15 ml reaction tubes and pelleted by centrifugation at 3000 rpm (SX4250 rotor/ Beckman X-22R Benchtop Falcon Centrifuge) for 5 minutes at room temperature. The pellet was resuspended in 300 µl of 0.6 M sorbitol, 20 mM Hepes, pH 7.4, 80 mM KCl, 2 mM PMSF, transferred to a 1.5 ml reaction tube and 200 µl cold glass beads (diameter: 0.5 mm) were added. Cells were lysed by four-times mixing steps with a vortexer (VF2/Janke und Kunkel) for 30 sec. Samples were placed on ice in between. The supernatant after centrifugation at 1000 xg and 4 °C for 3 min was transferred to a fresh reaction tube and spun down a second time for 10 min at 14000 rpm (12154-H rotor/ Sigma) and 4 °C. The supernatant of this spin was transferred to a new reaction tube and proteins were precipitated with the TCA method (see *2.3.1.5*). Pellet fractions were directly resuspended in 20 µl 2x Laemmli buffer with 5 % (v/v) β-mercaptoethanol. Supernatant and pellet fraction, representing cytosolic and crude mitochondrial fractions, respectively, were subjected to SDS-PAGE, transferred to nitrocellulose membrane by western blot and analyzed by immuno-staining with antibodies against the proteins of interest and diverse mitochondrial and cytosolic marker proteins as controls.

2.4.4 "Big Mitoprep" (Modified after Daum *et al.*, 1982)

For large-scale mitochondrial isolation, yeast cells were grown for 2 - 3 days before inoculation of the main culture (2 - 8 l) to an OD_{600} of ca. 0.1 in the evening. The cells were constantly kept in the logarithmic phase and diluted when necessary. In case proteins were to be expressed from yeast expression plasmid pYES2, 0.5 % galactose was added 2 hours before Mitoprep at OD_{600} of 0.6 for induction of the *GAL* promoter of the plasmid. At OD_{600} of 0.6 - 0.8, the culture was collected in 1 l centrifugation beakers and spun down for 5 min at 4000 rpm (JLA-81000 rotor) and room temperature. The cell pellet was resuspended in H_2O and transferred to JA-10 beakers. After centrifugation (5 min, 4000 rpm/ JA-10 rotor, RT) and removal of supernatant, the wet weight of the cell pellet was measured. The cell pellet was resuspended in 2 ml buffer 1 (100 mM Tris, 10 mM DTT) per 1 g wet weight and incubated at 30 °C with shaking for for ten minutes. Subsequently, the cells were spun down (5 min, 4000 rpm, JA-10 rotor, RT) and the pellet was resuspended in 50 - 100 ml 1.2 M sorbitol. After centrifugation the cell pellet was resuspended in 6.6 ml buffer 2 (1.2 M sorbitol, 20 mM KH_2PO_4, pH 7) including 6 g Zymolyase 20T per 1 g wet weight and incubated at 30 °C in the shaker for 45 min.

Spheroplasts were harvested by centrifugation (5 min, 2500 rpm/ JA-10 rotor, 2 °C) and the pellet was resuspended thoroughly in 12.6 ml homogenization buffer (0.6 M sorbitol, 10 mM Tris, pH 7.4, 1 mM EDTA, 1 mM PMSF, 0.2 % BSA/ fatty acid-free) per 1 g wet weight. The suspension was transferred to a pre-cooled glass homogenizer (Kontes, New Jersey), 12x homogenized on ice and transferred to JA-10 tubes. Unopened cells were pelleted by centrifugation (5 min, 3300 rpm/ JA-10 rotor, 2 °C) and the supernatant of this centrifugation was again centrifuged. The supernatant was transferred to JA-10 tubes and centrifuged for 12 min at 10000 rpm (JA-10 rotor) and 2 °C. The supernatant was discarded and the mitochondrial pellet was resuspended in 30 ml SH buffer (0.6 M sorbitol, 20 mM Hepes, pH 7.4) and transferred to pre-cooled JA-20 tubes. After centrifugation (5 min, 4000 rpm/ JA-20 rotor, 2 °C) the supernatant was transferred to fresh JA20 beakers and again centrifuged at 4000 rpm (2 °C) for 5 min. The supernatant was then again transferred to new JA-20 tubes and mitochondria were pelleted by centrifugation at 12000 rpm (JA-20 rotor) and 2 °C for 12 min. The mitochondrial pellet was resuspended in a small amount of SH buffer adjusted to the size of the pellet. The protein concentration of the isolated mitochondria was determined with the Bradford assay (see *2.3.1.6*). Mitochondria were divided into aliquots of appropriate amounts, snap-frozen in liquid nitrogen and stored at -80 °C.

2.4.5 Generation of mitoplasts

For disruption up the outer mitochondrial membrane, 100 µg of isolated mitochondria were resuspended in swelling buffer (0.2 M Hepes, pH 7.4) with 1.5 mM ATP and with or without 0.5 mg/ml proteinase K (PK). After incubation on ice for 20 min, proteinase K digestion was stopped adding 1 mM PMSF, and samples were incubated on ice for a further 5 min. After centrifugation (14000 rpm/ 12154-H rotor/ Sigma, 10 min, 4 °C), pellets were resuspended in 20 µl 2 x Laemmli buffer with 5 % (v/v) β-mercaptoethanol, heated at 95 °C for 5 min, subjected to SDS-PAGE and western blot and analyzed by immuno-staining with antibodies against the protein of interest and marker proteins for the outer membrane, intermembrane space, inner membrane and matrix.

2.4.6 Digitonin fractionation of mitochondria

Isolated mitochondria were diluted in SH (0.6 M sorbitol, 20 mM Hepes, pH 7.4) to a final concentration of 1 µg/µl and increasing amounts of 10 % digitonin solution were added to the following final concentrations: 0, 0.005, 0.01, 0.02, 0.03, 0.04, 0.05, 0.06, 0.08 and 0.1 %. In addition, proteinase K was added to the samples (50 µg/ml final concentration). One additional sample was kept in SH without digitonin and proteinase K as a control. Samples were kept on ice for 25 minutes, 200 µM PMSF was added to stop the proteinase K reaction and samples were incubated on ice for a further 5 minutes. After centrifugation for 10 min at 14000 rpm (12154-H rotor/ Sigma) and 4 °C, the pellet was resuspended in 2x Laemmli buffer with 5 % (v/v) β-mercaptoethanol, heated to 95 °C for 5 min and subjected to SDS-PAGE. Western blot nitrocellulose membranes were immuno-stained with antibodies against the protein of interest and different mitochondrial marker proteins of the outer membrane, intermembrane space, inner membrane and matrix.

2.4.7 Protease treatment

To investigate the folding states of DHFR constructs *in vivo*, isolated mitochondria were solubilized with 1 % Triton X-100 in the presence or absence of 20 µg/ml proteinase K. After incubation for 15 min on ice, 2 mM PMSF was added to stop the proteinase K reaction. After centrifugation at 14000 rpm (12154-H rotor/ Sigma) and 4 °C for 10 min, the pellet was dissolved in 2 x Laemmli buffer containing 5 % (v/v) β-mercaptoethanol, heated to 95 °C for 5 min and subjected to SDS-PAGE, western blot and immuno-staining for analysis. One set of samples was treated with the dihydrofolate substrate homolog methotrexate (200 µM final concentration) for 10 min prior to addition of Triton X-100 and PK. Samples were then analyzed by SDS-PAGE, western blot and immuno-staining.

2.4.8 Aggregation assay

In order to assess the aggregation of proteins expressed *in vivo*, isolated mitochondria were solubilized with 0.5 % Triton X-100, soluble and aggregate fractions were separated by centrifugation at 14000 (12154-H rotor/ Sigma) and 4 °C for 10 min and subsequently analyzed by SDS-PAGE, western blot and immuno-staining. Where indicated, isolated mitochondria were pre-incubated for 10 min at 25 °C, either in the presence of 1.5 mM ATP and 1.5 mM NADH to increase the mitochondrial ATP levels, or with 0.1 U/ml apyrase and 8 µM oligomycin to deplete ATP. Subsequently, mitochondria were incubated for 3 min at 25 °C or 42 °C prior to solubilization. Samples were then analyzed by SDS-PAGE, western blot and immuno-staining.

2.4.9 Ni-NTA agarose pulldown

Isolated mitochondria were incubated for 10 min at 25°C in the presence of 1 mM ADP or ATP or were depleted of nucleotides by addition of 0.1 U/ml apyrase, and 8 µM oligomycin. Subsequently, mitochondria were solubilized for 20 min at 4 °C with 1 % (w/v) digitonin in 20 mM Tris, pH 8.0, 80 mM KCl, 10 % glycerol, 20 mM imidazole, 1 mM PMSF. After centrifugation at 14000 rpm (12154-H rotor/ Sigma) and 4 °C for 10 min, solubilized material was mixed with Ni-NTA agarose beads (Qiagen) and incubated on an overhead roller for 60 min at 4°C. After three washing steps, proteins that were stably and specifically bound to the beads were eluted with Laemmli buffer containing 5 % (v/v) β-mercaptoethanol and 500 mM imidazole. Samples were analyzed by SDS-PAGE, western blot and immuno-staining.

2.5 Immunological methods

2.5.1 Overview of antibodies prepared during this thesis

Antibody	Antigen	Rabbit number
α-Cyc1	full length Cyc1 (Sigma-Aldrich; # C2436)	442 (animal facility, Pettenkoferstraße)
α-Imp1	MBP-Imp1 (full length)	470 (animal facility,

		Pettenkoferstraße)
α-Yme1	Yme1_AAA(250 - 525)-His6	474 (animal facility, Pettenkoferstraße)
α-Yme1	C-terminal peptide (CDERKDIGDDKPKIPTMLNA)	Tiere 1, 2, 3 (Pineda, Berlin)

2.5.2 Further antibodies used in this study

Antibody	Description
α-Aac2	C-terminal peptide with His6-tag, *S. cerevisiae*, Endres M.
α-Ccp1	*S. cerevisiae*, Langer T.
α-Cox2	C-terminal peptide, *S. cerevisiae*, Hell K.
α-Cyb2	Cyb2(168 - 557)-His$_6$, *S. cerevisiae*, Hell K.
α-DHFR	monoclonal, *M. musculus*, Origene
α-Dld1	C-terminal peptide, *S. cerevisiae*, Rojo E.
α-Erv1	Erv1-GST, *S. cerevisiae*, Herrmann H.
α-Fcj1	C-terminal peptide, *S. cerevisiae*, Rabl R.
α-Gut2	Gut2-His$_6$, *S. cerevisiae*, Harner M.
α-Hep1	Hep1(full length)-His$_6$, *S. cerevisiae*, Sichting M.
α-His	monoclonal, Qiagen
α-Hsp60	raised against *E. coli* GroEL, kind gift of Prof. Ulrich Hartl/ MPI, Martinsried
α-Mcs10	MBP-Mcs10-Afl, *S. cerevisiae*, Mokranjac D.
α-Mcs12/ -Aim5	MBP-Mcs12 (14 - 116), *S. cerevisiae*, Harner M.
α-Mcs19/ -Aim13	C-terminal peptide (EKSPSPQAKKTAIDK), *S. cerevisiae*, Harner M.
α-Mcs27/ -Aim37	C-terminal peptide (AKKCDLKRQIDQTLQ), *S. cerevisiae*, Harner M.,
α-Mge1	full length, *S. cerevisiae*, Sichting M.
α-Oxa1	N-terminal peptide, *S. cerevisiae*, Preuß M.
α-myc	monoclonal, *H. sapiens*, Sigma-Aldrich
α-Phb2	*S. cerevisiae*, Rabl R.

α-Ssc1	C-terminal peptide, *S. cerevisiae*, Baumann F.
α-Tim13	Tim13-MBP, *S. cerevisiae*, Terziyska N.
α-Tim14	GST-Tim14 (1 - 65), *S. cerevisiae*, Mokranjac D.
α-Tim17	C-terminal peptide, *S. cerevisiae*, Mokranjac D.
α-Tim21	His$_6$-Tim21 (97 - 239), *S. cerevisiae*, Mokranjac D.
α-Tim23	MBP-Tim23 (1 - 98), *S. cerevisiae*, Mokranjac D.
α-Tim44	Tim44 (43 - 431)-His$_6$, *S. cerevisiae*, Mokranjac D.
α-Tim50	CTim50-IMS domain, *S. cerevisiae*, Mokranjac D
α-Tom22	N-terminal peptide, *S. cerevisiae*, Harner M.
α-Tom40	C-terminal peptide, *S. cerevisiae*
α-Tom70	N-terminal peptide, *S. cerevisiae*, Käser M.

2.5.3 Generation of specific antisera in rabbits

Polyclonal antisera were generated in rabbits. Purified recombinant proteins expressed in *E. coli* were loaded onto SDS-PAGE and transferred to nitrocellulose membrane by western blot. The nitrocellulose membrane was stained with Ponceau S and the bands containing the protein of interest were cut out, destained and solubilized in DMSO by vortexing prior to subcutaneous injection into rabbit. For the first injection, 200 μg protein in a volume of 250 μl were mixed 1:1 with TiterMax reagent (Sigma-Aldrich) for 1 min on the vortexer. The emulsion was subcutaneously injected into the neck area of the rabbit in three steps. Further injections were performed monthly whereby the 200 μg protein in a 250 μl volume were mixed 1:1 with Freund's Adjuvans incomplete (Sigma-Aldrich). Ca. 10 days after each injection (except the first), 10 - 30 ml blood were taken from the vein of the ear. After coagulation during incubation at RT for 60 min, the blood sample was centrifuged twice (1st: 3000 x *g*, 5 min, RT; 2nd: 20000 x *g*, 15 min, RT) to separate the serum containing antibodies from the blood clot. The serum (supernatant) was transferred to a scintillation vial, incubated at 56 °C for 20 min for inactivation of the complement system and stored at -20 °C.

2.5.4 Detection of proteins on nitrocellulose membranes by immuno-staining

Upon transfer of the proteins onto nitrocellulose, unspecific binding sites were blocked by incubation of the membrane in milk blocking solution (5 % (w/v) milk powder (Spinnrad) in 1x TBS (150 mM NaCl, 10 mM Tris/HCl pH 7.5) or in 3 % (w/v) BSA for 30 min at RT on the shaker. Subsequently, the membrane was incubated in a solution of antiserum or purified antibody in blocking solution. Dilutions were chosen according to the strength of the antibody (1:200 - 1:500) or the antiserum (1:200 and 1:20000). After incubation for 2 h at RT or over night at 4 °C on the shaker, the membrane was washed in three 5 min steps on the shaker: 1x TBS, 1x TBS with 0.05 % TritonX-100, 1x TBS. Subsequently, the membrane was incubated in the appropriate dilution of monoclonal goat anti-rabbit (1:10000) or goat anti-mouse (1:5000) antibodies coupled to horseradish peroxidase (both Bio-Rad) in blocking solution. After shaking at RT for 1 h, the membrane was washed in three steps of 5 min as described before and incubated for 1 min in luminol reagent composed out of solution ECL1 and ECL2 in a ratio 1:1 (see table below). The membrane was then put between two transparent plastic foils and exposed to X-ray films for different time periods (typically 1sec, 5 sec, 15 sec, 30 sec, 1 min, 5 min).

ECL1 (30 ml final)	100 mM Tris/HCl, pH 8.5, 300 µl luminol (440 mg/10 ml DMSO), 132 µl p-cumaric acid (150 mg/10 ml DMSO)
ECL2 (30 ml final)	100 mM Tris/HCl, pH 8.5, 0.024 % (w/v) H_2O_2

2.6 Special Methods

2.6.1 Mass spectrometry of elution fractions of Ni-NTA agarose pulldown

In case the elution fractions of Ni-NTA agarose pulldown were to be analyzed by mass spectrometry, 12 mg of isolated mitochondria were applied. Otherwise, Ni-NTA pulldown was performed as decribed under 2.4.9. The gel was then stained with Coomassie and whole lanes were manually excised and digested with trypsin at 30 °C overnight (Shevchenko et al., 2000; Wilm et al., 1996). In the Imhof laboraty, peptides from trypsin digest were loaded into an Ultimate 3000 HPLC system (LC Packings) as described (Forné et al., 2012). The outflow from the HPLC was injected into an LTQ-Orbitrap mass spectrometer by electrospray method (Thermo Fisher Scientific). The survey full scan MS spectra (m/z 300 - 2000) were obtained with resolution R = 60,000 at m/z 400 after accumulation to target value (500000) in the linear ion trap. The six most intense peptide ions charged from 2 to 4 were sequentially isolated to a target value of 10000 and fragmented by collision induced dissociation (CID). They were collected and detected in the linear ion trap. For all Orbitrap measurements, 3 lock-mass ions were used for internal calibration (Olsen et al., 2005). The MS conditions were 1.5 kV spray voltage, 200 °C capillary temperature, 35 % normalized CID energy, activation q = 0.25, 30 msec activation time. Proteins were identified using Mascot version 2.3.02 (Database, Swissprot 57.10; Taxonomy, *S. cerevisiae*; MS tol, 10 ppm; MS/MS tol, 0.5 Da; peptide FDR, 0.1; protein FDR, 0.01; minimal peptide length, 5; variable modifications, oxidation (M); fixed modifications, carbamidomethyl). The fold-change of the proteins was quantified by applying spectral counting (Scaffold 3.0.9.1). Proteins that were used for quantification were filtered using 95.0 % peptide thresholds, 99.0 % protein thresholds and a minimum of 2 peptides. Proteins that showed at least five-fold enrichment in pulldown from mitochondria containing IMS-DHFR^{WT}-His over mitochondria containing matrix-DHFR^{WT}-His or the empty plasmid were considered as potential interaction partners of IMS-DHFR^{WT}-His.

2.6.2 Identification of aggregating proteins by SILAC and mass spectrometry

Wild-type YPH499 and *Δymel* strains were grown in medium containing normal and heavy lysine (Lys8: $^{13}C_6^{15}N_2$-L-Lysine•HCl, $^{13}C_6H_{14}^{15}N_2O_2$). After isolation of mitochondria, ATP was depleted by adding 0.1 U/ml apyrase and 8 µM oligomycin and an aggregation assay was performed as described above. The aggregate fractions of wild-type and *Δymel* mitochondria grown on heavy and light lysine, respectively, were pooled after the aggregation assay. In the other set of samples, aggregates from wild-type mitochondria isolated from cells grown on light lysine were mixed with aggregates from mitochondria isolated from *Δymel* cells grown on heavy lysine. Proteins were separated by SDS-PAGE and stained in Coomassi. After in-gel digestion with Lys-C protease at 30 °C over night, mass spectrometric analysis was performed as described in 2.6.1.

Maxquant (version 1.0.13.13) was used in combination with Mascot (version 2.3.02) for identification and quantification of proteins. The conditions for Mascot protein identification were as described above. The conditions for Maxquant were as follows: peptides for protein quantification = 1, unique and razor = 1, 1 peptide minimum, 2 min ratio count, multiplicity = 2, heavy labels = Lys8. Mitochondrial proteins that showed in at least two out of four independent experiments a 1.6-fold or higher aggregation in mitochondria lacking Yme1 compared to wild-type mitochondria (SILAC ratios of at least two different peptides) were included and considered as potential substrate candidates for Yme1.

2.7 Chemicals, consumables and equipment

2.7.1 Chemicals

Compound	Source/ Company
Acrylamide-Bisacrylamide	Serva
Agarose (for electrophoresis)	Serva
Acetic acid ($C_2H_4O_2$)	AppliChem
Adenine	Sigma-Aldrich
Amino acids (His, Lys, Leu, Trp)	Sigma-Aldrich
Ampicillin	AppliChem
Adenosine diphosphate (ADP)	Roche
Adenosine triphosphate (ATP)	Roche
Apyrase	Sigma-Aldrich
Bacto-peptone	Becton-Dickinson
β-Mercaptoethanol	Sigma-Aldrich
Bradford Reagent	Bio-Rad
Bromophenol blue	Roth
BSA	Roth
BSA, fatty acid-free	Serva

CaCl$_2$ (calcium chloride)	Merck
Citric acid (C$_6$H$_8$O$_7$)	Serva
Coomassi Brilliant Blue, CBB R-250	Serva
Cytochrome c	Sigma-Aldrich
Creatine kinase (CK)	Roche
Creatine phosphate (CP)	Sigma-Aldrich
P-Cumaric acid	Sigma-Aldrich
Dimethyl sulfoxide (DMSO)	Roth
Dithiotreitol (DTT)	Roth
dNTPs	New England Biolabs
Dry milk powder	Spinnrad
Ethanol	Sigma-Aldrich
Ethidium bromide	Serva
Freund's Adjuvans (incomplete)	Sigma-Aldrich
D-Galactose	Sigma-Aldrich
Gel extraction kit (DNA)	Promega
Glycerol	Sigma-Aldrich
Goat Anti-Mouse / -Rabbit	Bio-Rad
H$_2$O$_2$	AppliChem
Hepes	Serva
Hydrochloric acid (HCl)	AppliChem
Imidazole	AppliChem
IPTG (isopropyl β-D-1-thiogalactopyranoside)	Serva/ Merck
KCl (potassium chloride)	Merck
KH$_2$PO$_4$ (potassium dihydrogen phosphate)	Merck
Lactic acid (C3H6O3)	Merck
Lithium acetate	Sigma-Aldrich
L-Lysine-^{13}C$_6$, ^{15}N$_2$ hydrochloride	Sigma-Aldrich
Luminol	Sigma-Aldrich
Lysozyme	Serva
Milk powder	Spinnrad
MgCl$_2$ (magnesium chloride)	Merck

MgSO$_4$ (magnesium sulfate)	AppliChem
MnCl$_2$ (manganese chloride)	Merck
NaCl (sodium chloride)	Merck
NADH (nicotinamide adenine dinucleotide)	Serva
NaOH (sodium hydroxide)	ThermoFisher Scientific
Na$_2$HPO$_4$ (di-sodium hydrogen phosphate)	Merck
NaO$_2$C$_2$H$_3$ (sodium acetate)	Merck
NaO3 (sodium nitrate)	Serva
N$_2$O$_8$S$_2$H$_8$ (ammonium peroxo disulfate)	Serva
NH$_4$Cl (ammonium chloride)	Merck
(NH$_4$)$_2$SO$_4$ (ammonium sulfate)	Merck
Ni-NTA agarose	Qiagen
Oligomycin	Sigma-Aldrich
Pfu polymerase	New England Biolabs
PMSF (phenylmethansulfonyl fluoride)	Serva
Ponceau S solution	Sigma-Aldrich
Protease Arrest Reagent	Calbiochem
Proteinase K (PK)	Roche
Restriction enzymes	New England Biolabs
RNasin	New England Biolabs
Sodium dodecyl sulfate (SDS)	Serva
D-Sorbitol	Sigma-Aldrich
S^{35}-methionine	MP Biomedicals
SulfoLink Resin	Pierce/ ThermoFisherScientific
T4 DNA ligase	New England Biolabs
T4 ligation buffer (10x)	New England Biolabs
Taq polymerase	New England Biolabs
TEMED (tetramethylethylenediamine)	Serva
Titermax Gold	Sigma-Aldrich
Tris, buffer grade	AppliChem
Triton X-100	Sigma-Aldrich
Unstained Protein Marker	Fermentas/ ThermoFisherScientific

Uracil	Sigma-Aldrich
Urea	Serva
Valinomycin	Sigma-Aldrich
Yeast extract	Serva
Yeast nitrogen base	Invitrogen/ Life technologies
Zymolyase 20T	Seikagaku Biobusiness

2.7.2 Consumables

Item	Source/ Company
Blotting paper	NeoLab
Filters (sterile)	Schleicher & Schüll
Gloves (Latex, Nitrile)	Semper care
Nitrocellulose membrane	Whatman
PCR tubes	Peqlab
Petri dishes	Sarstedt
Pipette tips (10 - 1000 µl)	Sarstedt
1.5 ml reaction tubes	Sarstedt
2 ml reaction tubes	Sarstedt
15 ml reaction tubes	Greiner
50 ml reaction tubes	Greiner
Semi micro-cuvettes	Greiner
Super RX films	Fujifilm
X-ray films Biomax MR	Kodak

2.7.3 Equipment

Item	Source/ Company
Accu-jet	Brand
Balance CP2202S	Sartorius
Balance XS205	Mettler Toledo
Centrifuge Avanti J-25/ J-20 XP	Beckman
Centrifuge Benchtop Allegra X-22R	Beckman

MATERIAL AND METHODS

Centrifuge Table Top 5415D	Eppendorf
Centrifuge Ultracentrifuge optima MAX-E	Beckman
Developing machine CURIX60	AGFA
Electroporation system "Gene Pulser"	Bio-Rad
Gel imaging system "GelDoc-It"	UVP
Horizontal rotary shaker	made in house
PCR cycler "Mastercycler gradient"	Eppendorf
Pipettes, 10 µl - 5000 µl	Gilson
Sonifier 250	Branson
Spectrophotometer Nanodrop 2000c	Peqlab
Thermomixer "comfort"	Eppendorf

3. RESULTS

3.1 Generation and characterization of model substrates for investigating folding in the mitochondrial intermembrane space

3.1.1 Generation of an IMS-targeted Cytochrome b_2-DHFR model substrate

A well studied mouse protein was chosen as model substrate for the investigations. *Mus musculus* DHFR is essential for nucleotide biosynthesis in mouse. It catalyzes the reduction of dihydrofolic acid to tetrahydrofolic acid, using NADPH as an electron donor (Gangjee *et al.*, 2009). Mature DHFR attains a very compact structure that can be further stabilized by the substrate analogue methotrexate (Junker *et al.*, 2005). DHFR is frequently used to study import into and folding in isolated yeast mitochondria (Eilers and Schatz, 1986; Pfanner *et al.*, 1987; Gruhler *et al.*, 1995; Kanamori *et al.*, 1997; Gaume *et al.*, 1998).

DHFR was fused behind a targeting sequence derived from the endogenous yeast protein cytochrome b_2 (Cyb_2). A bipartite presequence targets Cyb_2 to the mitochondrial intermembrane space (Beasley *et al.*, 1993). The typical N-terminal matrix-targeting signal directs the precursor of Cyb_2 emerging from cytosolic ribosomes to the translocases of the outer and inner mitochondrial membrane, the TOM complex and the TIM23 complex. A stop-transfer signal is located C-terminally to the matrix-targeting signal and leads to the arrest of translocation at the level of the inner membrane. Subsequently, TIM23 opens laterally and releases the precursor protein into the lipid bilayer of the inner mitochondrial membrane. Matrix processing peptidase (MPP) cleaves off the matrix targeting signal (Yang *et al.*, 1988) and the mitochondrial inner membrane protease (IMP) cleaves the precursor at the intermembrane space side of the inner membrane (Pratje and Guiard, 1986; Geissler *et al.*, 2000). Consequently, the mature Cyb_2 is released as soluble protein into the intermembrane space. DHFR was fused to amino acid residues 1-107 of the Cyb_2 precursor including the bipartite presequence via a short glycine linker. The resulting construct was named IMS-DHFR^{WT} (Fig. 5). In order to generate a matrix-targeted control protein, the stop-transfer signal was deleted from

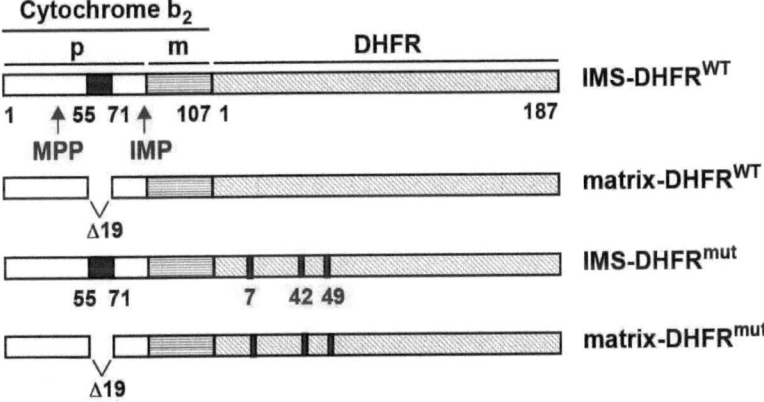

Figure 5. Schematic representation of the model substrates
Wild-type (WT) and non-folding mutant (mut; C7S, S42C, N49C) versions of mouse DHFR were fused C-terminally to amino acid residues 1-107 of yeast cytochrome b_2 targeting it to the intermembrane space (IMS-DHFR). Constructs with deleted stop-transfer signal should target to the matrix (matrix-DHFR). p, bipartite targeting signal of cytochrome b_2; cleavage by the mitochondrial processing peptidase (MPP) leads to the intermediate sized i-form; cleavage by the inner membrane peptidase (IMP) leads to the mature m-form.

the Cyb_2 presequence (1-107; Δ19). The resulting hybrid protein, matrix-DHFR^{WT}, should not be arrested at the level of the inner membrane but fully translocated into the matrix. Two further model substrates containing a mutant version of DHFR were created. Three point mutations (S7C/C42S/C49N) generate a folding incapable version of DHFR (Vestweber and Schatz, 1988). This mutant version of DHFR was fused to both Cyb_2 presequences, generating the two mutant hybrid proteins IMS-DHFR^{mut} and matrix-DHFR^{mut}. The inserts coding for the four model substrates (Fig. 5) were cloned into the yeast expression vector pYES2 and transformed into YPH499 wild-type yeast strain. Expression of the model substrates was induced by activation of the *GAL* promoter on the pYES2 plasmid. The expected submitochondrial localization of the model substrates is depicted in Figure 6.

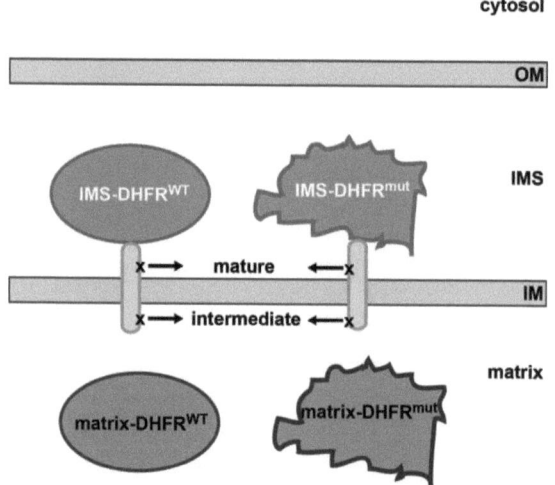

Figure 6. Overview of the predicted submitochondrial localization of the model substrates
Wild-type (WT) and non-folding mutant (mut) versions of DHFR were fused to Cyb_2(1-107) or Cyb_2(1-107; Δ19) to target them to the intermembrane space, IMS (red) and the matrix (blue), respectively.

3.1.2 Optimization of induction of model substrate expression

The time point for induction of the expression was then optimized. For this purpose, yeast cells harboring IMS-DHFR^{WT} or matrix-DHFR^{WT} on the pYES2 plasmid were grown for three to four doubling times in the logarithmic phase. Subsequently, one aliquot of cells was induced in the logarithmic growth phase (OD_{600} 0.6) and a second aliquot was induced in the stationary growth phase (OD_{600} 0.9). Samples of the cells before and after induction were withdrawn, solubilized with NaOH and the expression levels of the DHFR constructs were analyzed by SDS-PAGE, western blot, and immuno-staining. The efficiency of induction was very low when the promoter was activated in the stationary phase (Fig. 7, lanes 3+4, 7+8). Additionally, a significant background expression occurring without activation of the promoter was observed when cells were grown in the stationary phase. In contrast, during logarithmic growth, no background expression was detected and induction of the expression occurred with high efficiency (Fig. 7, lanes 1+2, 5+6).

Therefore, in further experiments, cells were kept in the logarithmic growth phase (OD$_{600}$ 0.6) and expression of the model substrates was also induced in this growth phase.

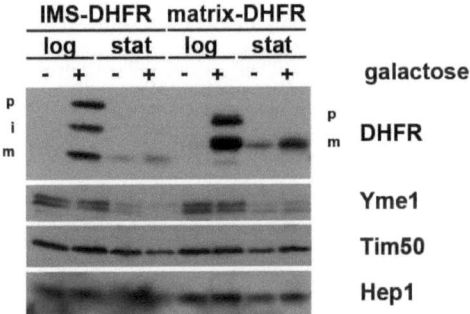

Figure 7. Optimization of model substrate expression
Two sets of yeast cells harboring IMS- or matrix-DHFR^{WT} in pYES2 plasmid were grown in parallel and expression was induced in the logarithmic (log) phase (OD$_{600}$ 0.6) in one set and in the stationary (OD$_{600}$ 0.9) phase (stat) in the other set by activation of *GAL* promoter. Cells were harvested after two hours induction, disrupted with NaOH and expression levels were analyzed by immuno-staining with antibodies for indicated proteins. p, precursor, i, intermediate and m, mature forms of the DHFR constructs.

3.1.3 Expression of model substrate in *S. cerevisiae*

Expression of the four model substrates was induced by activation of the *GAL* promoter and monitored over 120 minutes. Samples of the cultures were taken at different time points and total cell extracts were analyzed by SDS-PAGE, western blot and immuno-staining. The expression levels of the model substrates increased during time after induction (Fig. 8 A-D). The expression levels of the wild-type model substrates were slightly higher than the expression levels of the mutant model substrates (Fig 8 A, D). Moreover, the expression levels of matrix-DHFR were generally higher than in the intermembrane space (Fig. 8 C, D). In conclusion, these findings confirm that the model substrates can be stably expressed upon activation of the promoter and that the expression increases with time after activation. The lower expression levels of the folding incapable mutant DHFR constructs may be caused by their increased degradation. The lower expression levels in the intermembrane space could be attributed to a higher turnover of the overexpressed model substrates in this compartment. The volumetric capacity of the intermembrane space is low and therefore, this increased turnover of the overexpressed model substrates could prevent their aggregation. Unrelated yeast proteins of the cytosol (hexokinase) and the inner membrane (Tim17) were analyzed as controls. Neither of them was affected by the expression of the model substrates. In further experiments, expression of the model substrates was always induced for 120 minutes.

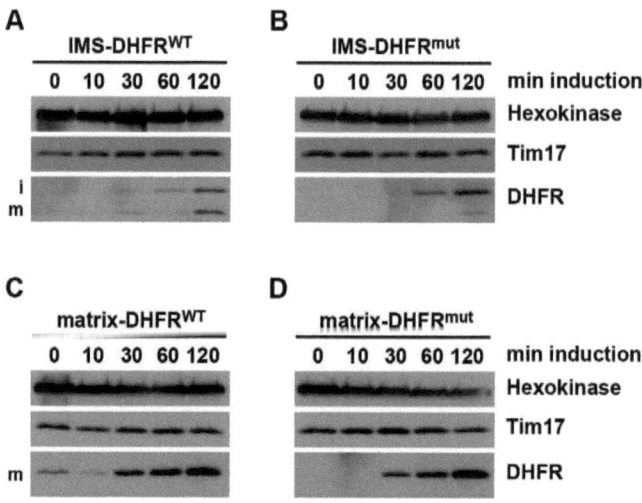

Figure 8. Kinetics of model substrate expression
Expression of model substrate was induced in the logarithmic growth phase and samples of cells taken at indicated time points. Cells were disrupted by the Rödel's cell lysis and the expression levels of IMS-DHFR (A, B) and matrix DHFR (C, D) constructs were analyzed by immuno-staining for indicated proteins. i, intermediate and m, mature form of IMS-DHFR constructs.

3.1.4 Verification of the steady state levels of endogenous proteins in cells expressing the model substrates

Possible side effects of the expression of the model substrates on the integrity of mitochondria were investigated. For this purpose, the steady state levels of a number of unrelated proteins from the four mitochondrial subcompartments were analyzed in mitochondria isolated from cells expressing the model substrates. None of the tested proteins of the outer membrane (Tom22), the intermembrane space (Cyb_2), the inner membrane (Tim50) and the matrix (Ssc1) were affected by the expression of the model substrates (Fig. 9). The expression levels of the endogenous proteins were indistinguishable among the four types of mitochondria.

In conclusion, this shows that all four mitochondrial subcompartments remain evenutally unaffected by expression of the model substrates.

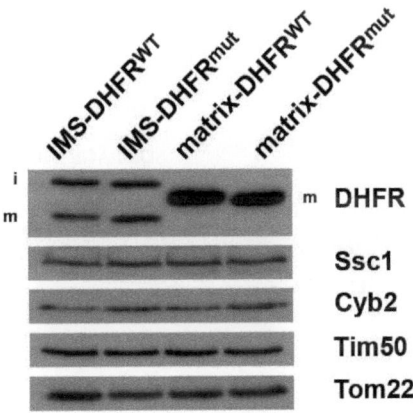

Figure 9. Steady-state levels of endogenous mitochondrial proteins in cells expressing the model substrates
Mitochondria were isolated and analyzed by SDS-PAGE, western blot and immuno-staining using antibodies against the indicated proteins. i, intermediate and m, mature form of IMS-DHFR constructs.

3.1.5 Subcellular localization of model substrates expressed *in vivo*

In order to analyze the subcellular localization the model substrates, fast mitochondrial isolation was performed from cells expressing IMS-DHFR^{WT} or matrix-DHFR^{WT}. Unbroken cells, cell debris and nuclei were removed and the crude mitochondrial fraction (M) was separated from the crude cytosolic fraction (C). The four model substrates were only detected in the crude mitochondrial fraction similar to the mitochondrial marker proteins Tom40 and Tim17 (Fig. 10, lanes 1+3). In contrast, the cytosolic marker protein Bmh2 was only detected in the cytosolic fraction (Fig. 10, lanes 2+4).

In summary, this shows that the model substrates are translocated into mitochondria and do not accumulate in the cytosol.

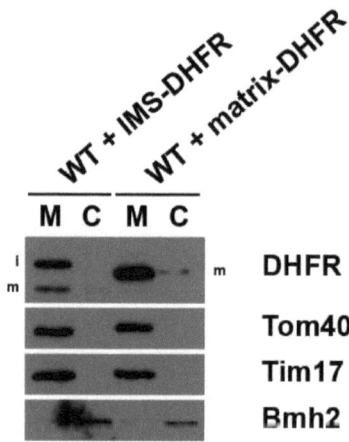

Figure 10. Subcellular localization of the DHFR constructs
Crude mitochondrial fraction (M) and remaining cellular components (C) from cells expressing the model substrates were analyzed by SDS-PAGE and immunostaining using antibodies against the indicated proteins. i, intermediate and m, mature form of IMS-DHFR.

3.1.6 Submitochondrial localization of the model substrates

The submitochondrial localization of the model substrates expressed in *S. cerevisiae* was determined by digitonin fractionation. The outer and the inner mitochondrial membrane were opened sequentially by increasing concentrations of digitonin as judged by the accessibility of marker proteins of the outer membrane (Tom70), the inner membrane (Tim50) and the matrix (Hep1). Externally added protease, proteinase K (PK), degraded outer membrane proteins first and then proteins of the intermembrane space and the inner membrane. Eventually, the matrix proteins were degraded upon disruption of the inner membrane. IMS-DHFR^{WT} and IMS-DHFR^{mut} behaved similar to Tim50, an inner membrane protein with a large domain protruding into the intermembrane space (Fig. 11A, C). This finding confirms that both IMS-targeted model substrates are indeed localized in the intermembrane space. Similarly to the matrix protein Hep1, matrix-DHFR^{WT} and matrix-DHFR^{mut} were

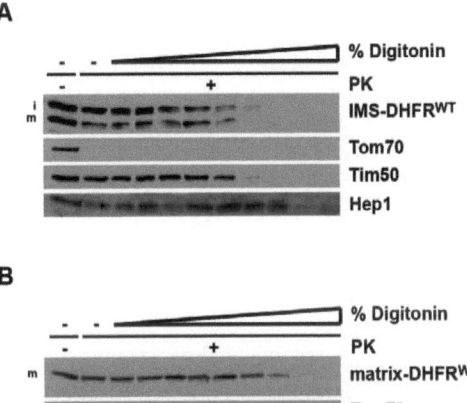
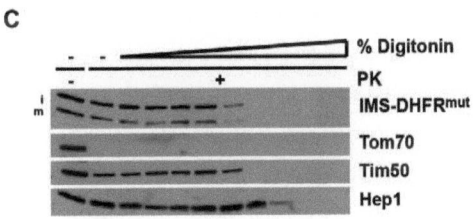
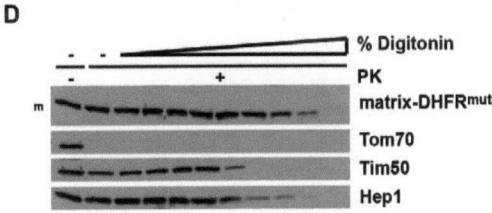

Figure 11. Submitochondrial localization of model substrates
Mitochondria were incubated with increasing amounts of digitonin (0.005 – 0.1 %) in the presence of proteinase K (PK) for 25 min on ice. Samples were analyzed by SDS-PAGE, western blot and immuno-staining using antibodies against DHFR and the indicated mitochondrial marker proteins (Tom70, outer membrane; Tim50, intermembrane space; Hep1, matrix). i, intermediate and m, mature form of IMS-DHFR constructs. Localization of wild-type (A, B) and mutant (C, D) constructs.

protected against proteinase K digestion until the inner membrane was disrupted. This finding demonstrates that the two matrix-targeted model substrates are also localized correctly (Fig. 11B, D).

Collectively, these findings show that all four DHFR constructs were sorted to the expected mitochondrial subcompartment.

3.2 Investigation of the folding behavior of the model substrates

3.2.1 *In vivo* protease resistance in the absence and presence of methotrexate

Protease treatment of DHFR had previously been shown to yield a stable fragment of roughly 25 kDa (Gaume *et al.*, 1998). In order to determine the folding state of *in vivo* expressed model substrates, their stability was tested upon protease treatment. For this purpose, isolated mitochondria from cells expressing the model substrates were solubilized and treated with proteinase K. The wild-type model substrates were partly degraded upon addition of the protease and a protease-resistant fragment of roughly 25 kDa was generated (Fig. 12, left panel, lane 2). In contrast, the mutant versions of the model substrates were entirely degraded (Fig. 12, right panel, lane 2) (Vestweber and Schatz, 1988). In presence of the DHFR substrate analog methotrexate, the wild-type model substrates were stabilized even further, and thus the levels of the stable fragment increased (Fig. 12, left panel, lane 3). This finding can be explained by binding of methotrexate to wild-type DHFR, which in turn leads to further stabilization of the native fold. Neither of the mutant forms was stabilized by methotrexate (Fig. 12, right panel, lane 3) because the substrate analog cannot bind to unfolded DHFR.

In conclusion, these results show that the wild-type DHFR constructs assume their native fold in the mitochondrial intermembrane space and in the matrix.

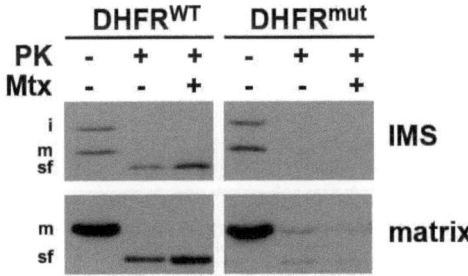

Figure 12. Folding of model substrates in the presence and absence of methotrexate
Isolated mitochondria were solubilized with Triton X-100 and incubated with proteinase K (PK) for 20 min at 0 °C in the presence and absence of methotrexate (Mtx). Samples were analyzed by SDS-PAGE and immuno-staining with antibodies against DHFR. i, intermediate and m, mature form of IMS-DHFR. sf, stable fragment upon protease digestion.

3.2.2 Requirements for folding of DHFR in the IMS and matrix

Next, it was asked if nucleotides and heat stress have an effect on the folding state of the DHFR constructs. To this end, one set of isolated mitochondria from cells expressing the model substrates was depleted of nucleotides. The ATP levels of a second set of these mitochondria were kept high. Both sets were subjected to a short heat shock at 42 °C for 3 min. One set of samples was kept at 25 °C as a control. After solubilization of the mitochondria with Triton X-100, pellet and supernatant fractions, representing aggregated and soluble proteins, were separated by centrifugation. At 25 °C, IMS-DHFR^{WT} was found in the soluble fraction in the presence and absence of ATP (Fig. 13 A, lanes 1-4). However, upon heat shock, the mature (m) form of IMS-DHFR^{WT} aggregated in an ATP-dependent manner (Fig. 13 A, lanes 5-8). In the presence of ATP, 14 % of IMS-DHFR^{WT} aggregated (Fig. 13 A, lane 7). In the absence of ATP, aggregation increased to 89 % (Fig. 13 A, lane 5). The intermediate (i) form of IMS-DHFR^{WT}, which is not yet cleaved by inner membrane peptidase and is thus still anchored to the inner membrane, aggregated almost completely upon heat shock (99 %) (Fig. 13 A, lanes 5+7).

Matrix-DHFR^{WT} also aggregated in an ATP-dependent manner upon heat shock (Fig. 13 B, lanes 5-7). However, only a much smaller proportion (1 % or 3 %) than in the intermembrane space aggregated in the matrix upon heat shock (Fig. 13 B, lanes

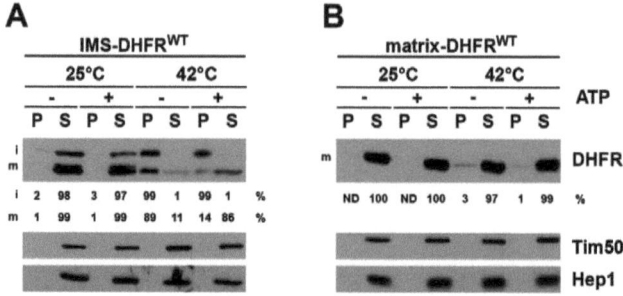

Figure 13. Aggregation of wild-type DHFR constructs in IMS and matrix
Isolated mitochondria were incubated under the indicated conditions, solubilized with Triton X-100, and soluble (S) and aggregate (P, pellet) fractions were separated by centrifugation and analyzed by SDS-PAGE and immuno-staining for indicated proteins. Singals of wild-type IMS-DHFR (A) and matrix-DHFR (B) were quantified in supernatant and pellet fractions and expressed as percentages of total protein. ND, not detectable. i, intermediate and m, mature form of IMS-DHFR.

5+7). This could be attributed to a higher capacity of the protein quality control systems of the matrix than of the ones in the intermembrane space.

IMS-DHFR^{mut} aggregated under all tested conditions (Fig. 14 A), even in the presence of ATP under physiological conditions (Fig. 14 A, lane 3). Matrix-DHFR^{mut} behaved similar to matrix-DHFR^{WT} (Fig. 14 B), although aggregation upon heat shock was slightly increased in comparison to the wild-type counterpart (Fig. 14 B, lanes 5+7). In the presence of ATP, only 10 % of matrix-DHFR^{WT} aggregated upon heat shock (Fig. 4, lane 7). This is in agreement with the ATP-dependent activity of the chaperones of the protein quality control system in the mitochondrial matrix.

Collectively, these results show that, under stress conditions, the stability of DHFR is determined by an ATP-dependent process. This applies to the wild-type constructs in the intermembrane space and in the matrix. Furthermore, it becomes clear that the capacity of the protein quality control system in the matrix is higher than the capacity of the protein quality control system in the intermembrane space. The unfolded DHFR^{mut} construct was kept soluble in the matrix whereas its counterpart in the intermembrane space aggregated even under physiological conditions.

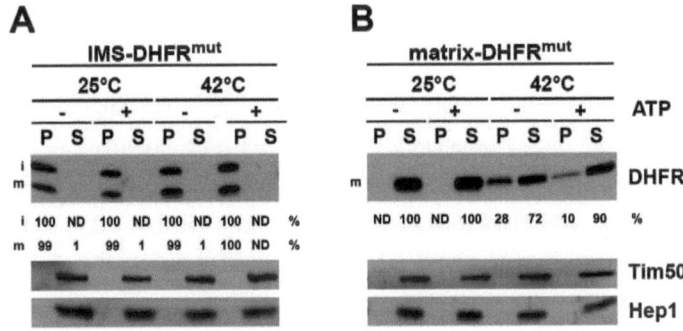

Figure 14. Aggregation of DHFR^{mut} in IMS and matrix
Isolated mitochondria were incubated under conditions to increase or decrease the mitochondrial ATP levels and then exposed to 25°C or 42°C for three minutes. Mitochondria were then solubilized with Triton X-100-containing buffer and soluble (S) and aggregate (P, pellet) fractions separated by centrifugation and analyzed by SDS-PAGE and immuno-staining using antibodies against the indicated proteins. The DHFR signals were quantified in the supernatant and pellet fractions and expressed as percentages of total protein. ND, not detectable. i, intermediate and m, mature form of IMS-DHFR.

3.3 Identification of potential folding helpers of DHFR in the IMS

3.3.1 Ni-NTA pulldown and label-free quantification by mass spectrometry

Having confirmed that wild-type DHFR assumes its native fold in the intermembrane space, it should be investigated, which folding helpers guide this process. For this purpose, the binding partners of DHFR in the intermembrane space needed to be determined. Yeast strains expressing a C-terminally His-tagged version of wild-type IMS- and matrix-DHFR were employed. Mitochondria isolated from cells expressing IMS-DHFR^{WT}-His or matrix-DHFR^{WT}-His were solubilized with digitonin and the His-tagged model substrates were captured with Ni-NTA agarose beads. Mitochondria harboring empty pYES2 vector served as a control. The eluate fractions from the agarose beads were subjected to SDS-PAGE, excised from the gel and analyzed by mass spectrometry (Fig. 15). Three proteins were specifically co-isolated with IMS-DHFR^{WT}-His: Yme1, Mgr1, and Mgr3. Yme1 is the AAA protease of the mitochondrial intermembrane space mediating the turnover of inner membrane and intermembrane space proteins (Nakai *et al.*, 1995; Pearce and Sherman, 1995; Weber *et al.*, 1995; Lemaire *et al.*, 2000; Kominsky *et al.*, 2002; Augustin *et al.*, 2005; Graef *et al.*, 2007; Osman *et al.*, 2009a; Tamura *et al.*, 2009; Potting *et al.*, 2010). Mgr1 and Mgr3 have recently been shown to form a supermolecular complex with Yme1 (Dunn et al., 2006; Dunn et al., 2008). Furthermore, Mgr1 and Mgr3 have been suggested to function as adaptors of Yme1 (Dunn et al., 2008; Graef *et al.*, 2007; Osman *et al.*, 2009; Potting *et al.*, 2010).

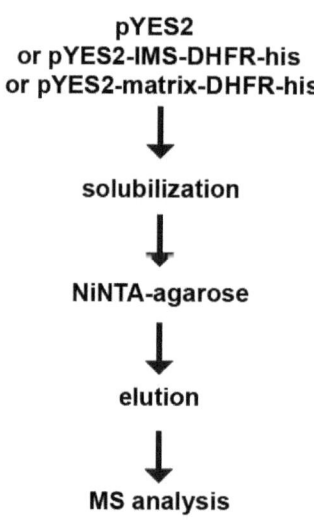

Figure 15. Workflow of Ni-NTA-pulldown and mass spectrometry
Upon Ni-NTA pulldown and isolation of His-tagged DHFR constructs, samples were separated by SDS-PAGE. Proteins in the elution fractions were excised, digested with trypsin and tryptic fragments were quantitatively analyzed by label-free mass spectrometry (MS).

3.3.2 Confirmation of Ni-NTA pulldown by western blot and immuno-staining

The results of the mass spectrometric analysis were confirmed with an unrelated method. For this purpose, Ni-NTA pulldown was performed as described above. This time, mitochondria were pretreated with ATP-depleting system, ATP-regenerating system or ADP prior to solubilization. Samples of the mitochondrial lysates (total) and the eluates (bound) were analyzed by SDS-PAGE, western blot and immuno-staining. Yme1 was specifically co-isolated with IMS-DHFR^{WT}-His (Fig. 16, lanes 11, 14), but not with matrix-DHFR^{WT}-His (Fig. 16, lanes 12, 15, 18). Notably, Yme1

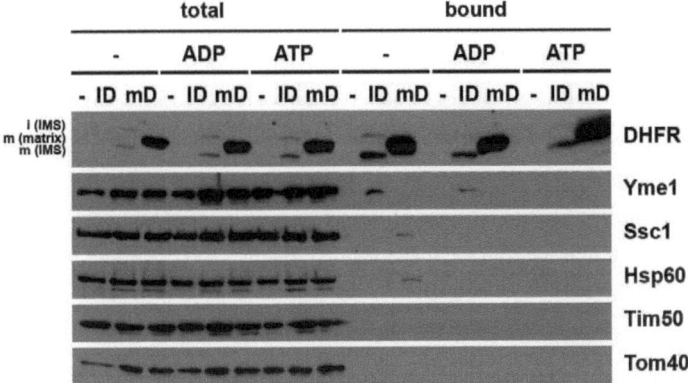

Figure 16. Identification of potential folding helpers of IMS-DHFR
Isolated mitochondria were solubilized with digitonin-containing buffer in the absence of nucleotides or in the presence of ATP or ADP. Samples were incubated with Ni-NTA agarose beads and specifically bound proteins eluted with Laemmli buffer containing 500mM imidazol. Total (10 %), and bound (100 %) fractions were analyzed by SDS-PAGE and immuno-staining for the indicated proteins. ID, IMS-DHFR, mD, matrix-DHFR, i, intermediate and m, mature form of IMS-DHFR.

was only co-isolated in the absence of nucleotides or in the presence of ADP (Fig. 16, 'ID' in '-/ADP-bound'). In the presence of ATP, Yme1 was not co-isolated (Fig. 16, 'ID' in 'ATP/bound'). These findings agree with the typical behaviour of chaperone-substrate interactions. The interaction is typically stabilized by ADP or in the absence of nucleotides, as the substrate can bind to the chaperone but is not processed. In the presence of ATP, the interaction is shorter as the substrate is processed and released very quickly. Therefore, the interaction is not easy to detect. Importantly, the two matrix chaperones Hsp70 (Ssc1) and Hsp60 were specifically co-isolated with matrix-DHFR^{WT}-His in the absence of ATP. This is in line with former studies reporting that Hsp70 and Hsp60 are involved in the folding of DHFR in the matrix (Gaume et al., 1998; Ostermann et al., 1989).

Collectively, these experiments demonstrate an ATP-dependent interaction of IMS-DHFR^{WT} with Yme1. This suggests that Yme1 could function as a chaperone for DHFR in the intermembrane space.

3.4 Generation and characterization of Yme1 deletion strain

3.4.1 Growth phenotype of Δyme1 strain

To analyze a potential chaperone-like function of Yme1, *YME1* was deleted in the YPH499 background and the growth behavior of the deletion strain was analyzed. The *YME1* locus was replaced by a kanamycin cassette using homologous recombination. The growth behavior of the deletion strain was analyzed at different temperatures on fermentable (YPD) and non-fermentable (YPG) medium. A general growth defect of Δ*yme1* strain was observed on non-fermentable medium, irrespective of the growth temperature (Fig. 17, left panel). The deletion of Yme1 induced a temperature-sensitive growth defect on fermentable medium at high (37 °C) and low (14 °C) temperature (Fig. 17 right panel). These observations differ slightly from the findings of previous studies on the *YME1* deletion strain. These studies report growth defect only on non-fermentable medium at 37 °C and on fermentable medium at 14 °C (Thorsness *et al.*, 1993; Thorsness and Fox, 1993). The likely reason for this discrepancy is that in the present study a different background strain for the *YME1* deletion. There are studies confirming that the background strain can affect the phenotype of particular deletions to a large extent (Dunn *et al.*, 2008).

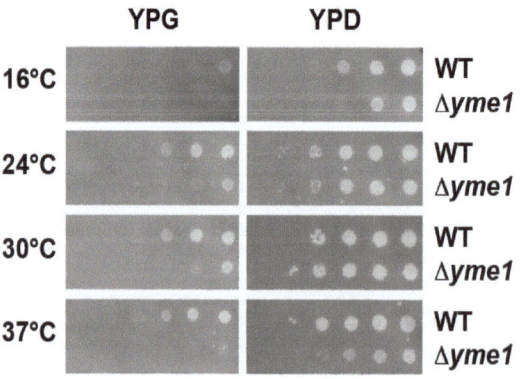

Figure 17. Growth phenotype of Δ*yme1*
Wild-type and Δ*yme1* strain were grown for three days in YPGal. Five serial dilutions of cell suspension were prepared. 2 µl of each dilution were spotted on YPG and YPD plates. The plates were incubated at the indicated temperatures. WT, wild-type.

3.4.2 Mitochondrial DNA in Δ*ymel* strain

In order to analyze the presence of mitochondrial DNA in the Δ*ymel* strain, PCR was performed on the mitochondrially encoded gene *ATP8*. Δ*hep1*, a strain (Burri *et al.*, 2004)lacking mitochondrial DNA (rho^0) was analyzed as a control. The PCR fragment of *ATP8* was detected in both wild-type and Δ*ymel* strains but not in the Δ*hep1* strain (Fig. 18). In conclusion, this result demonstrates that the Δ*ymel* strain does not lose its entire mitochondrial genome although an elevated rate of mitochondrial DNA escape to the nucleus had been reported in previous studies (Thorsness *et al.*, 1993; Thorsness and Fox, 1993).

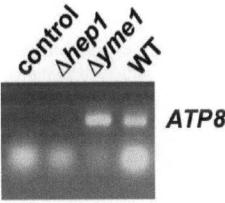

Figure 18. Mitochondrial DNA of Δ*ymel*
PCR on mitochondrially encoded *ATP8* was performed in WT, Δ*ymel* and Δ*hep1* strains. Amplified DNA was visualized under UV light. water control, control. WT, wild-type.

3.5 Behavior of the model substrate in the absence of Ymel

3.5.1 Expression of model substrates in Δ*ymel* strain

To address the role of Ymel in folding of DHFR in the IMS, IMS-DHFR^{WT} and matrix-DHFR^{WT} were expressed in the Δ*ymel* strain. The steady state levels of the model substrates were analyzed in isolated mitochondria by SDS-PAGE, western blot and immuno-staining. Interestingly, the steady levels of IMS-DHFR^{WT} were approximately two-fold higher in the Δ*ymel* strain than in the wild-type strain (Fig. 19, lanes 1-4). The expression levels of matrix-DHFR^{WT}, however, were indistinguishable between deletion and wild-type strain (Fig 19, lanes 5-8). This is in agreement with previous studies which show that DHFR is a proteolytic substrate of Ymel (Leonhard *et al.*, 1999), and supports the idea of a specific role of Ymel in the biogenesis of IMS-DHFR.

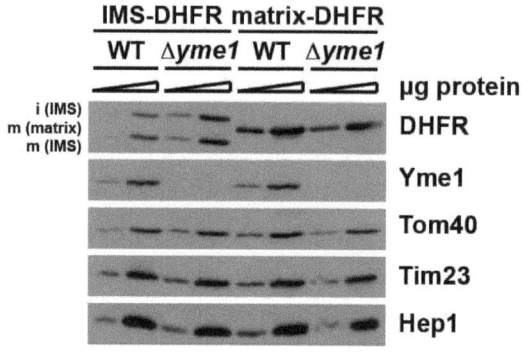

Figure 19. Expression of DHFR constructs in Δ*yme1*
Steady-state levels of isolated mitochondria (5 and 15 μg) from cells expressing the model substrates were analyzed by SDS-PAGE and immuno-staining for indicated proteins. i, intermediate and m, mature form of IMS-DHFR. WT, wild-type.

3.5.2 Effect of Yme1 on the folding of DHFR

Folding of IMS-DHFR^{WT} was analyzed in mitochondria lacking Yme1. For this purpose, isolated mitochondria were depleted from ATP, solubilized and soluble and aggregate fractions were separated by centrifugation. In the absence of Yme1, a large proportion of IMS-DHFR^{WT} was found in the aggregate fraction even under physiological conditions (Fig. 20 A, left panel, lane 1). This finding underlines that Yme1 is not only involved in degradation of DHFR, but likely also plays a role in folding DHFR in the intermembrane space. Interestingly, the intermediate form of IMS-DHFR^{WT} that is processed only once shows a stronger dependancy on Yme1 than the mature form (Fig. 20 A, left panel, lane 1). A possible explanation for this observation is that the membrane-anchor of the intermediate form slows down the folding of the mature part of the protein. Therefore, the intermediate form needs more intensive chaperone assistance. Importantly, the deletion of *YME1* had no effect on the folding of matrix-DHFR^{WT} (Fig. 20 B, left panel, lanes 1+2). Matrix-DHFR^{WT} was only detected in the soluble fraction in the presence and absence of Yme1 (Fig. 20 B, both panels).

In conclusion, these experiments demonstrate that Yme1 is not only involved in degradation of DHFR but strongly suggests a role of Yme1 in the folding of DHFR in the intermembrane space.

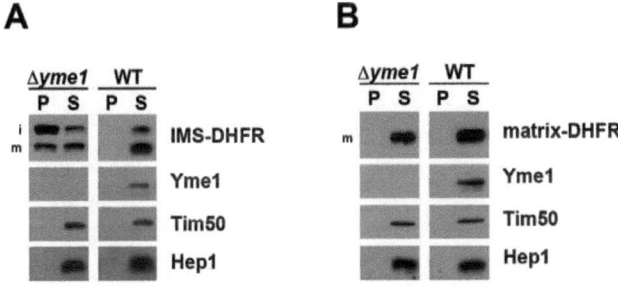

Figure 20. Aggregation of DHFR constructs in the *YME1* deletion strain
Mitochondria of cells expressing IMS-DHFR^{WT} (A) or matrix-DHFR^{WT} (B) were solubilized in Triton X-100-containing buffer and soluble, S, and aggregate, P (pellet), fractions were separated by centrifugation. Samples were analyzed by SDS-PAGE followed by western blot and immuno-staining for indicated proteins. i, intermediate and m, mature form of IMS-DHFR. WT, wild-type.

3.6 Identification of endogenous Yme1 substrates

3.6.1 Identification of proteins that aggregate in the absence of Yme1 by SILAC and mass spectrometry

Next, endogenous substrates of the chaperone-like activity of Yme1 should be determined and analyzed. For this purpose, wild-type and *Δyme1* cells were grown in medium containing heavy or light lysine (Fig. 21). Mitochondria were isolated, aggregate fractions were collected and mixed from both strains of mitochondria in a way that one strain is labeled with heavy lysine and the other strain is labeled with light lysine (Fig. 21). The combined aggregate fractions were subjected to SDS-PAGE and proteins were digested 'in gel' with the protease Lys-C. In collaboration with the Imhof laboratory (Zentrum für Proteinanalytik, Munich), the peptides resulting from Lys-C treatment were subsequently analyzed by mass spectrometry (Fig. 21).

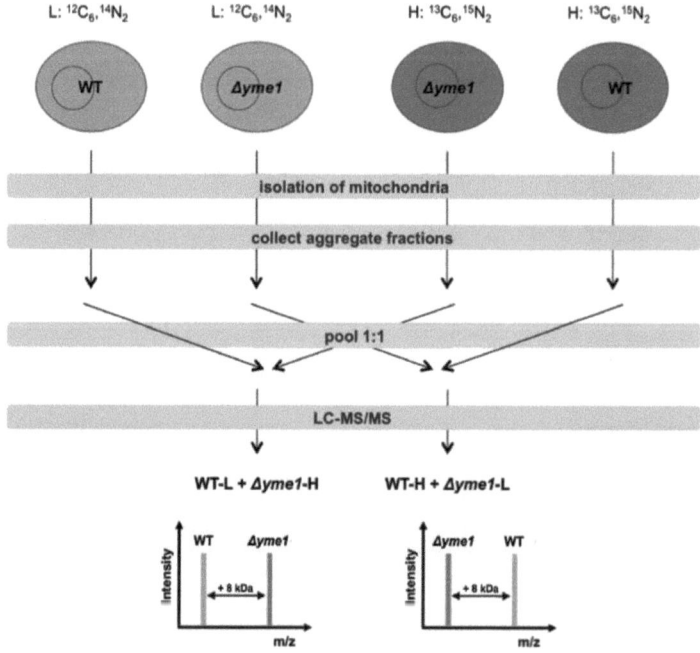

Figure 21. SILAC and mass spectrometry setup - a schematic overview
Wild-type and Δ*yme1* strain were grown in medium containing heavy or light lysine. Mitochondria were isolated, aggregate and soluble fractions separated, mixed from both types of mitochondria and subjected to SDS-PAGE. Peptides resulting from Lys-C digestion of gel slices from SDS-PAGE were analyzed by mass spectrometry.

Two biological and four technical replicates were analyzed and proteins that showed at least 1.6-fold higher aggregation in mitochondria lacking Yme1 than in mitochondria from the wild-type strain were classified as potential substrate proteins of the chaperone-like activity of Yme1. The list of mitochondrial proteins that fulfilled these criteria is given in Table 1. Among the identified potential Yme1 substrates are predominantly proteins that reside in the intermembrane space or the inner membrane. Among them were Ups2, an intermembrane space protein that is involved in mitochondrial lipid metabolism (Osman *et al.*, 2009; Tamura *et al.*, 2009;

Tamura *et al.*, 2012), three components of the dehydrogenase complex of the intermembrane space, Nde1, Dld1 and Gut2 (Augustin et al., 2005; Grandier-Vazeille et al., 2001), prohibitins, protein scaffolds of the inner membrane (Osman *et al.*, 2009; Steglich *et al.*, 1999) and Cox2, an essential component of the respiratory chain in the inner membrane (Nakai et al. 1995; Pearce and Sherman, 1995; Taanman and Capaldi, 1992; Weber *et al.*, 1996). Ups2, Nde1 and Cox2 were previously shown to represent the proteolytical substrates of Yme1 (Nakai *et al.*, 1995; Dunn *et al.*, 2008; Potting *et al.*, 2010). Most of the proteins of the mitochondrial matrix that were identified in the SILAC screen belong to the chaperone family, and the intrinsic aggregation propensity of members of this protein family is significantly higher compared to non-chaperone proteins. This could explain why these matrix-located proteins were found in the screen.

Notably, the identified proteins of the intermembrane space and the inner membrane are not uniform. They belong to different structural and functional classes. This observation suggests that Yme1 is not only important for a subset of proteins. It instead seems to have broad substrate specificity. The fact that the identified proteins aggregate in the absence of Yme1 indicates that Yme1 is not only involved in the degradation of these substrates, but also has a crucial function in their folding. The identification of previously known substrates confirms the method and strongly suggests that additional identified proteins are also true Yme1 substrates.

Table1. Proteins that aggregate in *Δyme1* mitochondria

Wild-type and *Δyme1* cells were grown in medium containing light or heavy lysine. Mitochondria were isolated and soluble and aggregate fractions separated by centrifugation as described before. After SDS-PAGE, aggregate fractions were excised from the gel and loaded on Orbitrap mass spectrometer. Two biological replicates were analyzed. Table 1 contains alphabetically sorted mitochondrial proteins that showed at least 1.6 fold higher aggregation propensity in mitochondria lacking Yme1 in at least two of the experiments. Only proteins with at least two different quantified peptides were considered. SL, submitochondrial localization; TM, transmembrane domains; CF, cofactor; SL/TM/CF according to the Uniprot database (www.uniprot.org); function according to the SGD database (www.yeastgenome.org).

	ORF	Protein	SL	TM	CF	Function
1	Q0250	COX2	IM-IMS	2	copper	subunit II of cytochrome c oxidase, mitochondrially-encoded
2	YAL039C	CYC3	IMS	-	heme, iron	cytochrome c heme lyase (holocytochrome c synthase), attaches heme to apo-cytochrome c in the IMS
3	YBL095W	-	not known	1	-	unknown
4	YBR262C	AIM5	IM-IMS	1	-	subunit of mitochondrial IM organizing system (MitOS/ MICOS/ MINOS), role in maintenance of crista junctions and IM architecture
5	YBR282W	MRPL27	matrix	-	-	mitochondrial protein of the large ribosomal subunit
6	YCL044C	MGR1	IM-IMS	2	-	subunit of mitochondrial i-AAA protease that degrades misfolded mitochondrial proteins, binds to substrates to facilitate proteolysis, required for growth of rho^0 cells
7	YCR071C	IMG2	matrix	-	-	mitochondrial ribosomal protein of the large subunit
8	YDL174C	DLD1	IM-IMS	1	FAD, zinc	D-lactate dehydrogenase, oxidizes D-lactate to pyruvate

9	YDR316W	OMS1	IM-IMS	1	-	with conserved methyltransferase motif; multicopy suppressor of respiratory defects caused by OXA1 mutations
10	YFL036W	RPO41	matrix	-	-	RNA polymerase; enhancing DNA bending and melting to facilitate pre-initiation open complex formation
11	YFR011C	AIM13	IMS	-	-	subunit of mitochondrial IM organizing system (MitOS/ MICOS/ MINOS), role in maintenance of crista junctions and IM architecture
12	YGL057C	GEP7	IM	1	-	unknown function; null mutant exhibits respiratory growth defect and synthetic interactions with prohibitin (Phb1) and Gem1
13	YGL068W	MNP1	matrix	-	-	protein associated with mitochondrial nucleoid, required for normal respiratory growth
14	YGR029W	ERV1	IMS	-	FAD	flavin-linked sulfhydryl oxidase, oxidizes Mia40p as part of the disulfide relay system
15	YGR076C	MRPL25	matrix	-	-	mitochondrial protein of the large ribosomal subunit
16	YGR132C	PHB1	IM-IMS	1	-	inner mitochondrial membrane chaperone that stabilizes newly synthesized proteins
17	YGR174C	CBP4	IM-IMS	1	-	required for assembly of cytochrome bc1 complex; interacts with the Cbp3-Cbp6 complex and newly synthesized cytochrome b to promote assembly of cytochrome b into cytochrome bc1 complex
18	YGR286C	BIO2	matrix	-	iron, sulfur	biotin synthase, catalyzes the conversion of dethiobiotin to biotin

19	YHL021C	AIM17	not known	-	iron	unknown; null mutant displays reduced frequency of mitochondrial genome loss
20	YHR005C	TIM10	IMS	-	zinc	essential IMS protein, forms a complex with Tim9 that delivers hydrophobic proteins to TIM22 complex for insertion into the IM
21	YHR024C	MAS2	matrix	-	zinc	large subunit of mitochondrial processing protease (MPP), essential processing enzyme, cleaves the N-terminal targeting sequences from mitochondrially imported proteins
22	YIL155C	GUT2	IM-IMS	1	FAD	mitochondrial glycerol-3-phosphate dehydrogenase
23	YJL066C	MPM1	not known	-	-	unknown function, no hydrophobic stretches
24	YJR045C	SSC1	matrix	-	ATP	Hsp70 family ATPase, constituent of the import motor component of TIM23 complex, involved in protein translocation and folding
25	YJR048W	CYC1	IMS	-	heme, iron	electron carrier of mitochondrial intermembrane space, transfers electrons from ubiquinone-cytochrome c oxidoreductase to cytochrome c oxidase during cellular respiration
26	YJR100C	AIM25	not known	-	-	unknown function, null mutant viable/ displays elevated rate of mitochondrial genome loss
27	YKL138C	MRPL31	matrix	-	-	mitochondrial ribosomal protein of the large subunit
28	YKL150W	MCR1	OM/IMS	1/-	FAD, NAD	mitochondrial NADH-cytochrome b5 reductase, involved in ergosterol biosynthesis

29	YKR016W	FCJ1	IM-IMS	1	-	ortholog of mammalian mitofilin, essential role in maintenance of crista junctions and IM architecture, component of mitochondrial IM organizing system (MICOS/ MitOS/ MINOS)
30	YLL027W	ISA1	matrix	-	-	required for maturation of mitochondrial (4Fe-4S) proteins
31	YLR168C	UPS2	IMS	-	-	role in regulation of phospholipid metabolism by inhibiting conversion of phosphatidylethanolamine to phosphatidylcholine
32	YLR203C	MSS51	matrix	-	-	translational activator of mitochondrial COX1 mRNA; influences COX1 mRNA translation and Cox1 assembly into cytochrome c oxidase
33	YML025C	YML6	matrix	-	-	mitochondrial protein of the large ribosomal subunit
34	YMR115W	MGR3	IM-IMS	1	-	subunit of mitochondrial i-AAA protease which degrades misfolded mitochondrial proteins, binds to substrates to facilitate proteolysis, required for growth of rho^0 cells
35	YMR145C	NDE1	IM-IMS	1	FAD, NAD	mitochondrial external NADH dehydrogenase, catalyzes oxidation of cytosolic NADH providing it to the respiratory chain
36	YMR203W	TOM40	OM-IMS	β-barrel	-	component of the TOM complex, responsible for recognition and initial import steps for all mitochondrially directed proteins
37	YNL100W	AIM37	IM-IMS	2	-	subunit of mitochondrial IM organizing system (MitOS/ MICOS/ MINOS), role in maintenance of crista junctions and IM architecture

38	YNR018W	RCF2	IM-IMS	2	-	cytochrome c oxidase subunit; role in assembly of respiratory supercomplexes; required for late-stage assembly of the Cox12 and Cox13 and for cytochrome c oxidase activity
39	YNR020C	ATP23	IMS	-	zinc	metalloprotease of the IM, required for processing of Atp6; role in assembly of the F0 sector of the F1F0 ATP synthase complex
40	YOR020C	HSP10	matrix	-	-	matrix co-chaperonin that inhibits the ATPase activity of Hsp60; involved in protein folding and sorting in mitochondria; similarity to *E. coli* groES
41	YOR211C	MGM1	IM-IMS/IMS	1/-	GTP	GTPase; complex with Ugo1 and Fzo1; required for mitochondrial morphology and genome maintenance; long and short form; homolog of human OPA1 involved in autosomal dominant optic atrophy

3.6.2 Endogenous levels of Yme1 substrates in *Δyme1* strain

The findings from the SILAC screen were confirmed by an independent method. For this purpose, the steady state levels of candidate proteins from the SILAC screen for which antibodies were available were determined in the *Δyme1* and wild-type strains. The endogenous levels of Erv1, Mcs10, and Mcs27 were approximately two-fold higher in the absence of Yme1 (Fig. 22, left panel). This is remniscent of the behavior of DHFR in the absence of Yme1, and suggests that these proteins are also proteolytically turned over by Yme1. The steady-state levels of the other substrate candidates from the SILAC screen, Mcs19, Fcj1, Phb2, Gut2, and Dld1, were indistinguishable between wild-type and deletion strains (Fig. 22, left and middle panel), indicating that these proteins are not predominantly degraded by Yme1. The steady-state levels of a large number of other mitochondrial proteins residing in the four mitochondrial subcompartments, outer membrane, intermembrane space, inner membrane and matrix, were tested and none of them was affected in the absence of Yme1 (Fig. 22, middle and right panel).

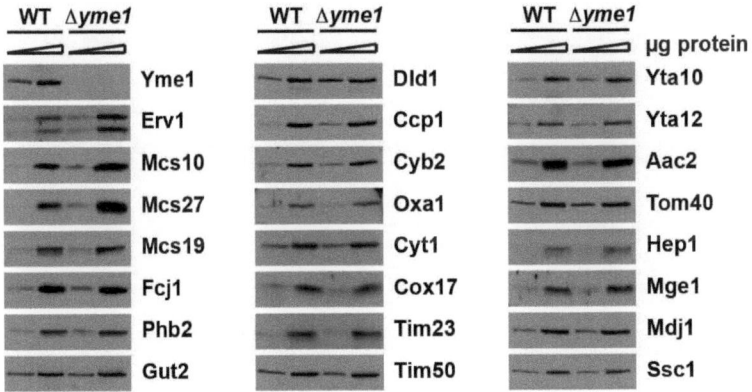

Figure 22. Steady state levels of endogenous proteins aggregated in *Δyme1*
Endogenous levels of candidates from SILAC were analyzed in wild-type and *Δyme1* mitochondria. Isolated mitochondria (5 and 15µg) were analyzed by SDS-PAGE, western blot, and immuno-staining with antibodies against indicated proteins. WT, wild-type.

3.6.3 Aggregation of endogenous Yme1 substrates in mitochondria of *Δyme1* strain

In order to confirm the data form the SILAC screen, the aggregation assay was repeated and aggregate and soluble fractions were analyzed by SDS-PAGE, western blot and immuno-staining with antibodies against the substrate candidates from the SILAC screen. Importantly, Erv1, Phb2, Gut2, Mcs19 and Dld1 were indeed found in the aggregate fraction of *Δyme1* but not of wild-type mitochondria (Fig. 23, left panel, lanes 1+3). The aggregation propensity of these proteins was even increased by a short heat shock at 42 °C for 3 min. Interestingly, Mcs27 and Fcj1, two components of the recently identified mitochondrial contact site complex MICOS/MINOS/MitOS (Harner et al., 2011, Hoppins et al., 2011, von der Malsburg et al., 2011) aggregated

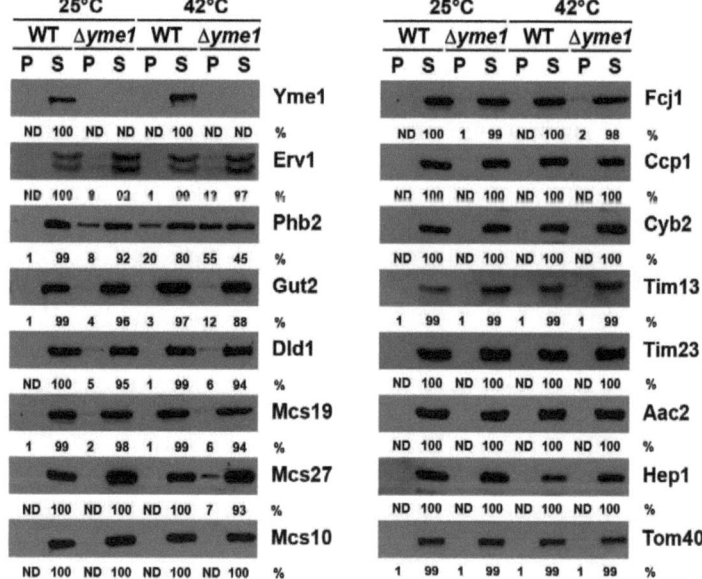

Figure 23. Aggregation of endogenous substrates in *Δyme1*
Mitochondria were pre-incubated for three minutes at 25 or 42 °C, solubilized with Triton X-100-containing buffer and soluble, S and aggregate, P (pellet) fractions separated by centrifugation. Samples were analyzed by SDS-PAGE followed by immuno-staining with the indicated antibodies. The DHFR signals were quantified in supernatant and pellet fractions and expressed as percentages of total protein. ND, not detectable.

in the absence of Yme1 only upon heat shock. This observation suggests that the chaperone-like function of Yme1 becomes more important under stress conditions. Notably, the *in vitro* folding of the sulfhydryl oxidase Erv1 was recently shown to depend on the introduction of disulfide bonds by Mia40 (Kallergi *et al.*, 2012). My data, however, suggest that an additional player is involved in the folding of Erv1: Yme1. The elevated endogenous levels of Erv1 and its aggregation in the absence of Yme1 indicate that Yme1 exerts a dual function. In the case of Erv1, Yme1 can switch between the proteolytic and the chaperone function. Aggregation of Erv1 in the absence of Yme1 could affect import of the substrates of the Mia-Erv1 disulfide relay system (Hell, 2008). Similarly, the aggregation of the intermembrane space dehydrogenases Dld1 and Gut2 could have secondary effects on the oxidative phosphorylation and thus respiratory growth (Grandier-Vazeille *et al.*, 2011). And finally, aggregation of components of the MICOS/MINOS/MitOS complex (Mcs19, Mcs27, Fcj1) could affect the mitochondrial ultrastructure that is coordinated by this complex (Harner *et al.*, 2011, Hoppins *et al.*, 2011, von der Malsburg *et al.*, 2011).

Collectively, these data confirm that Yme1 has a role in folding of a multitude of structurally and functionally distinct proteins of the intermembrane space and the inner membrane. Aggregation of these substrate proteins in the absence of Yme1 would certainly affect their function and this in turn could lead to multiple secondary and tertiary effects. Considering the physiological functions of the identified Yme1 substrates, the aggregation of these proteins could explain the diverse aspects of the pleiotropic phenotype of Yme1. In addition, the data indicate that Yme1 has a broad substrate specificity that is not restricted to a particular subset of intermembrane space and inner membrane proteins.

3.6.4 Characterization of Mpm1

The candidate protein from the SILAC study that showed the highest aggregation in *Δyme1* is Mpm1. Almost nothing is known about this protein, neither its submitochondrial localization nor its function. However, Mpm1 was co-isolated with Mcs10, a core component of the recently identified MICOS/MINOS/MitOS complex (Harner et al., 2011, Hoppins et al., 20122, von der Malsburg et al., 2011). This complex is located at contact sites between the inner and outer mitochondrial membranes and plays a crucial role in maintenance of the mitochondrial morphology.

Interestingly, the mitochondrial morphology is also strongly affected in cells lacking Yme1 (Campbell et al., 1998; Stiburek et al., 2012). Therefore, I sought to analyze Mpm1 in more detail. For this purpose, I generated a yeast strain containing a chromosomally myc-tagged version of Mpm1. Digitonin fractionation and alkaline extraction were performed to elucidate the localization of Mpm1. During digitonin fractionation, Mpm1 was accessible to externally added proteinase K (PK) as soon as the outer membrane was disrupted similar to the behavior of the inner membrane protein Tim50 (Fig. 24 A). When alkaline extraction was performed, Mpm1 was found in the soluble fraction (S), similar to the soluble matrix protein Hep1 (Fig. 24 B). Both results are consistent with the prediction programs, which suggest that Mpm1 has no transmembrane segments. Taken together, these results indicate that Mpm1 is a soluble protein of the mitochondrial intermembrane space.

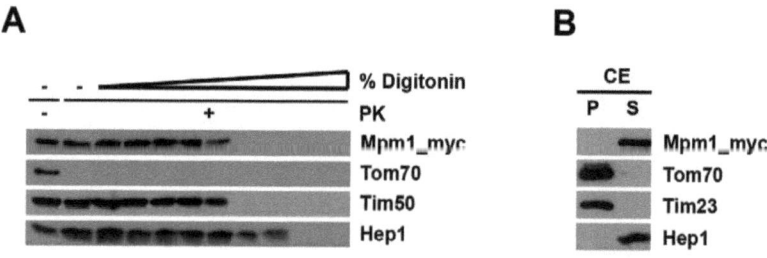

Figure 24. Submitochondrial localization of Mpm1
Mitochondria harboring the myc-tagged version of Mpm1 were subjected to digitonin fractionation as described in Figure 11 (A) and carbonate extraction, CE (B). Samples were analyzed by SDS-PAGE, western blot and immuno-staining for the indicated proteins. P, pellet and S, soluble fraction after carbonate extraction. PK, proteinase K.

3.6.5 Effect of deletion of *YME1* on Mpm1 expression levels

Next, the behavior of Mpm1 in absence of Yme1 should be analyzed. Analysis of solubilized isolated mitochondria from cells harboring chromosomally myc-tagged Mpm1 by SDS-PAGE, western blot and immuno-staining showed that the steady state levels of Mpm1 are indistinguishable between the wild-type and *YME1* deletion strains. This finding suggests that Mpm1 is not proteolytically turned over by Yme1 (Fig. 25).

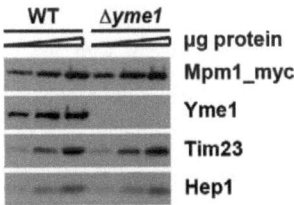

Figure 25. Steady state levels of Mpm1 in *Δyme1*
Isolated mitochondria (5, 15, and 30 µg) of cells expressing myc-tagged Mpm1 were solubilized, subjected to SDS-PAGE and western blot and analyzed by immuno-staining for indicated proteins. WT, wild-type.

3.6.6 Aggregation of Mpm1 in the absence of Yme1

Next, the aggregation assay was performed as described before with isolated mitochondria from cells harboring chromosomally myc-tagged Mpm1. The dependancy of the folding state of Mpm1 on the presence of Yme1 was clearly pronounced even under physiological conditions (Fig. 26, '25 °C'). Heat treatment at 42 °C did not enhance the aggregation of Mpm1 in the deletion strain (Fig. 26, '25 °C'). This result implies that the folding of Mpm1 is generally highly dependent on Yme1 and not only under stress conditions.

Taken together, the localization of Mpm1, its endogenous levels and its aggregation behavior in the absence of Yme1 indicate that Mpm1 is a *bona fide* chaperone substrate of Yme1.

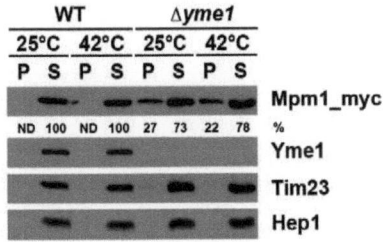

Figure 26. Aggregation of Mpm1 in *Δyme1*
Mitochondria of cells expressing myc-tagged Mpm1 were pre-incubated for three minutes at 25 or 42 °C, solubilized with Triton X-100-containing buffer and soluble, S and aggregate, P (pellet) fractions were separated by centrifugation. Samples were analyzed by SDS-PAGE, western blot and immuno-staining with the indicated antibodies. WT, wild-type.

3.6.7 Co-isolation of endogenous substrates of Yme1 with His-tagged Yme1

Finally, I wanted to determine if Yme1 interacts directly with the newly identified endogenous substrates. To answer this question, isolated mitochondria from a strain harboring N-terminally His-tagged Yme1 were pre-treated with ADP or ATP and solubilized with digitonin-containing buffer. His-tagged Yme1 was captured with Ni-NTA agarose beads and isolated. Samples of the mitochondrial lysate (total), the supernatant after capturing (flow through) and the eluates were subjected to SDS-PAGE, western blot and immuno-staining. The membrane was stained with antibodies against the newly identified Yme1 substrates.

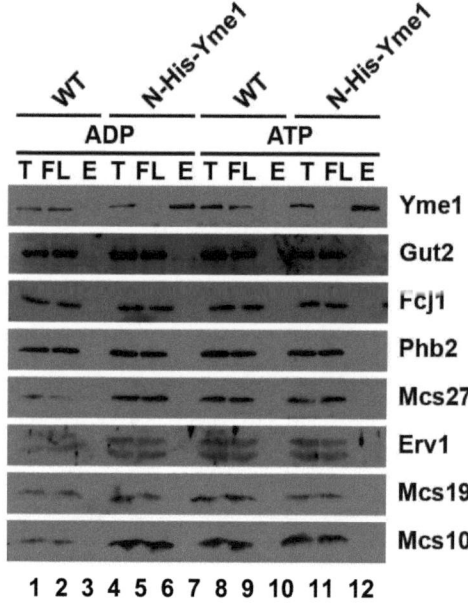

Figure 27. Co-isolation of Gut2 with His-tagged Yme1
Isolated mitochondria of cells expressing N-terminally His-tagged Yme1 were solubilized with digitonin-containing buffer in the presence of ATP or ADP. Samples were incubated with Ni-NTA agarose beads and specifically bound proteins eluted with Laemmli buffer containing 500 mM imidazol. Total (T, 10 %), flow through (FT, 10 %), and eluate (E, 100 %) fractions were analyzed by SDS-PAGE and immuno-staining with the indicated antibodies. WT, wild-type.

Gut2 was specifically co-isolated with His-tagged Yme1 in the presence of ADP but not in the presence of ATP (Fig. 27, lanes 6+12). As described before, this behavior is characteristic for most substrate-chaperone interactions. In the presence of ADP, the substrate binds to the chaperone but does not get processed. Therefore, the interaction between substrate and chaperone is longer than in the presence of ATP and is thus easier to detect by western blot and immuno-staining. None of the other tested newly identified Yme1 substrates was co-isolated with Yme1 under these conditions.

As discussed before, the endogenous levels of Gut2 were unaffected in absence of Yme1 but Gut2 aggregated in the *YME1* deletion strain. These findings and the confirmation of a direct interaction of Gut2 and Yme1 provide evidence that Gut2 is a substrate of the chaperone-like activity of Yme1.

4. DISCUSSION

4.1 Folding of the model substrate DHFR in the mitochondrial intermembrane space

Mitochondria originate from bacterial ancestors that were engulfed by a progenitor of eukaryotic cells (Gray, 1999). They are thus surrounded by two bilayers of protein-lipid membranes. In the course of time, a transfer of prokaryotic DNA to the eukaryotic nucleus took place. Therefore, nowadays, 99 % of mitochondrial proteins are encoded in the nucleus (Reichert and Neupert, 2004). A sophisticated translocation system evolved in order to transport these cytosolically synthesized proteins back to the organelle. Furthermore, the mitochondrial inner membrane and the matrix are entirely shielded from the cytosol and are not accessible to the cytosolic protein quality control systems. Thus, intrinsic protein quality control systems have developed to maintain mitochondrial protein homeostasis.

The majority of the components of the matrix-located protein quality control system are derived from the bacterial ancestor and exert similar functions as their homologs in today's bacteria and the eukaryotic cytosol (Sauer and Baker, 2011). These components have been studied in detail and structural and functional data are available. In contrast, very little is known about potential components or functions of a protein quality control system in the mitochondrial intermembrane space. Nevertheless, it is certain that such a system must exist. The mitochondrial IMS harbors many large and complex proteins and it is very unlikely that they assume their native fold without the assistance of molecular folding helpers (Herrmann and Neupert, 2003). The AAA protease Yme1 belongs to the proteolytic branch of the protein the quality control in the intermembrane space. It is involved in the turnover of soluble and membrane-anchored intermembrane space proteins.

One subset of relatively small proteins is imported and also folded by the recently discovered the Mia40-Erv1 disulfide relay system (Deponte and Hell, 2009; Koehler and Tienson, 2009; Herrmann and Riemer, 2012). Substrates of this pathway harbor conserved cysteine residues in the form of CX_3C or CX_9C motifs. Upon translocation through the TOM pore, the substrates interact with oxidized Mia40. It is assumed that the introduction of one or more disulfide bonds by Mia40 is sufficient to trigger folding of the entire substrate protein (Milenkovic et al., 2007; Banci et al., 2010).

In order to elucidate the folding processes in the mitochondrial intermembrane space, I targeted the frequently used model substrate mouse DHFR to the intermembrane space of yeast mitochondria. I fused DHFR behind the bipartite presequence of the endogenous intermembrane space protein cytochrome b_2 (IMS-DHFR^{WT}). The hybrid protein was then expressed *in vivo* in a wild-type yeast strain. The folding state of the model substrate was analyzed under physiological (25 °C) and stress (42 °C) conditions. Upon heat stress, the model substrate aggregated in the intermembrane space in a nucleotide dependent manner. In presence of ATP, the aggregation was significantly lower than without nucleotides. This observation hints at the presence of an ATP-dependent folding helper that prevents the model substrate from aggregation.

4.2 Role of Yme1 in folding of the model substrate DHFR

Next, a screen for potential folding helpers was performed. His-tagged IMS-DHFR^{WT} was pulled down and co-isolated proteins were identified by mass spectrometry. Yme1, the *i*-AAA protease of the mitochondrial inner membrane (Koppen and Langer, 2007), was found to specifically co-purify with the model substrate. Six Yme1 subunits form a homo-hexameric ring structure that is integrated into the inner membrane. The catalytic sites of Yme1 are exposed to the intermembrane space. Yme1 shows overlapping substrate specificity with its matrix-facing counterpart, *m*-AAA protease. Both proteins mediate the cooperative turnover of inner membrane proteins (Leonhard et al., 2000). Cox2 and Nde1 were the only confirmed proteolytic substrates of Yme1 for a long time (Nakai et al., 1995; Pearce and Sherman, 1995; Weber et al., 1996; Kominsky et al., 2002; Augustin et al., 2005; Graef et al., 2007; Dunn et al., 2008; Potting et al., 2010; Elliott et al., 2011). Only recently, diverse screening approaches in yeast revealed further proteolytic substrates. The rapid proteolytic turnover of Ups1 and Ups2, two proteins involved in the phospholipid metabolism of the inner mitochondrial membrane, is mediated by Yme1 (Osman et al., 2009a; Tamura et al., 2009; Potting et al., 2010; Tamura et al., 2012). Furthermore, it was recently shown that the prohibitins, Phb1 and Phb2, protein and

DISCUSSION

lipid scaffolds, are proteolytic substrates of Yme1 (Graef et al., 2007; Osman et al., 2009b). Already thirteen years ago, Leonhard et al. suggested a chaperone-like activity of Yme1 based on *in vitro* experiments (Leonhard et al., 1999). The isolated AAA domain of Yme1 was able to restore the activity of two model substrates, DHFR and rhodanese upon denaturation with urea. However, these *in vitro* studies have not been confirmed *in vivo* so far (Leonhard et al., 1999).

In the present *in vivo* study, it was observed that IMS-DHFR^{WT} aggregates in the absence of Yme1 even under physiological conditions. This finding strongly underlines the fact that Yme1 indeed has a chaperone-like function and that the model substrate in the intermembrane space depends on this function. Importantly, the two isoforms of IMS-DHFR^{WT}, behaved differently in the absence of Yme1. The mature form that is fully processed was distributed equally between soluble and aggregate fractions. However, the intermediate form that is not processed by the inner membrane protease was found predominantly in the aggregate fraction. This observation shows that the membrane-bound intermediate form, that is cleaved only once, depends to a greater extent on the presence of Yme1. The folding of membrane bound proteins may pose an even greater challenge to protein quality control systems than the folding of soluble proteins. In addition, Yme1 mediates proteolytic turnover of IMS-DHFR^{WT} substantiated by increased levels of the model substrate in the $\Delta yme1$ strain. Importantly, a recent *in vivo* study reported similar findings on a chaperone-like function of Yme1 (Fiumera et al., 2009). Fiumera et al. showed that Yme1 promotes the folding of Cox2 and probably assembly of Cox2 into the cytochrome c oxidase complex of the inner membrane.

In conclusion, Yme1 seems to monitor both processes: the folding and the proteolytic turnover of the model substrate IMS-DHFR^{WT}. It is conceivable that Yme1 mediates *de novo* folding of IMS-DHFR^{WT} upon import. Upon several unsuccessful folding attempts, Yme1 could then switch to the proteolytic mode and degrade the unfolded or misfolded DHFR construct. However, the question how Yme1 can "decide" between the proteolytic mode and the chaperone-like mode remains unanswered. Interestingly, a similar working mechanism has recently been hypothesized for AAA proteases in general (Sauer and Baker, 2011). Future studies are needed to answer this question conclusively.

4.3 Endogenous substrates of the chaperone-like activity of Yme1

In order to identify endogenous substrates of the chaperone-like activity of Yme1, I analyzed the endogenous proteins that aggregate in the absence of Yme1. Wild-type strain and $\Delta yme1$ strain were labeled with heavy and light isotopes of lysine in cell culture (SILAC). The aggregate fractions of wild-type and $\Delta yme1$ strain were thus distinguishable from one another and could be analyzed simultaneously by mass spectrometry. I identified a number of very diverse endogenous proteins that aggregated in the absence of Yme1. One of the aggregating proteins, Ups2, is a confirmed proteolytic susbtrate of Yme1. This indicates that Ups2 is not only turned over by Yme1 (Potting et al., 2010), but also dependent on Yme1 to assume and maintain its mature tertiary structure. The deletion of YME1 probably results in an impaired biogenesis of functional Ups2. The physiological function of Ups2 is in regulation of the phospholipid metabolism of the inner membrane (Osman et al., 2009b), and this could be the underlying mechanism causing an altered phospholipid composition of mitochondria in the $\Delta yme1$ strain (Nebauer et al., 2007). Furthermore, three components that are involved in cellular respiration, have been found to aggregate in the absence of Yme1: Dld1, Gut2 and Cox2, the confirmed proteolytic substrate of Yme1. Dld1 and Gut2 are dehydrogenases of the intermembrane space and Cox2 is a subunit of the cytochrome c oxidase complex in the inner membrane (Taanman and Capaldi, 1992; Khanday et al., 2002; Steinmetz et al., 2002; Lu et al., 2003). This could represent the underlying mechanism for the respiratory deficiency of YME1 deletion strain and the associated growth phenotype on non-fermentable carbon sources (Thorsness et al., 1993). As discussed before, Fiumera et al. also suggest that Yme1 is not only involved in the degradation of Cox2 (Fiumera et al., 2009). Their findings suggest that Yme1 chaperones the folding and/or assembly of Cox2. In contrast to the degradation of Cox2 by Yme1, this process seems to be independent of the two potential adaptor proteins Mgr1 and Mgr3.

Interestingly, Mpm1 has the highest aggregation propensity of all proteins identified in the $\Delta yme1$ strain in the SILAC approach. Yet, almost nothing is known about this mitochondrial protein. Inadome et al. studied the submitochondrial localization of Mpm1 but the results are contradictory (Inadome et al., 2001). In their study, Mpm1 was found in the membrane fraction after alkaline extraction. However, the primary sequence does not contain any segments that could function as

DISCUSSION

transmembrane domains. In a recent screen for proteins of the mitochondrial contact site complex, Mpm1 was detected amongst the soluble protein fraction (Harner M., personal communication). In the present study, I could confirm that Mpm1 is a soluble protein of the mitochondrial intermembrane space. This localization makes Mpm1 a suitable substrate for Yme1.

Importantly, in a recent study on the mitochondrial contact site complex, Mpm1 was co-isolated with Mcs10 (Hoppins *et al.*, 2011), a component of this complex, named MICOS/MINOS/MitOS (Harner *et al.*, 2011; Hoppins *et al.*, 2011; von der Malsburg *et al.*, 2011). MICOS links the inner boundary membrane and the outer mitochondrial membrane at cristae junctions (Harner *et al.*, 2011). In the present study, I identified the MICOS component Fcj1 in the SILAC screen for endogenous Yme1 substrates. Indeed, Fcj1 and two further components of MICOS, Mcs19 and Mcs27, were confirmed by western blot and immuno-staining to aggregate in the absence of Yme1. The deletion of two components of the MICOS complex, Fcj1 and Mcs10, leads to a strongly reduced amount of cristae junctions and mitochondrial branches (Harner *et al.*, 2011). This results in largely altered mitochondrial architecture as confirmed by electron microscopy. Similarly, the deletion of Yme1 also induces a severe defect in mitochondrial ultrastructure with grossly swollen and punctate mitochondria (Campbell *et al.*, 1994; Campbell and Thorsness, 1998). This phenotype is distinguishable from the structural changes observed in $\Delta fcj1$ or $\Delta mcs10$ strains. Nevertheless, both phenotypes could result from similar underlying mechanisms originating from changes in the composition of MICOS/MINOS/MitOS complex.

The prohibitins Phb1 and Phb2 also aggregate in the absence of Yme1. This confirms that prohibitins are not only proteolytic substrates of Yme1 (Osman *et al.*, 2009b). They also rely on the chaperone-like activity of the Yme1 to assume and/or maintain their mature fold. Prohibitins were shown to represent protein and lipid scaffolds. Their aggregation in the absence of Yme1 might cause a multitude of different phenotypes. Any aspect of the $\Delta yme1$ phenotype could be explained by the aggregation of the prohibitins. The aggregation of prohibitins could likely induce secondary effects on proteins that rely on Phb1 and Phb2 for their function or localization. Furthermore, prohibitins were shown to form a supercomplex with *m*-AAA protease in the inner membrane (Steglich *et al.*, 1999). Deletion studies revealed a regulatory function of prohibitins on the degradation of membrane proteins by *m*-AAA protease (Steglich *et al.*, 1999). Thus, aggregation of prohibitins in the absence of Yme1 might have profound effects on the activity of the *m*-AAA protease. The aggregation of the prohibitins could affect the activity of the *m*-AAA protease and this, in turn, would affect substrates of the *m*-AAA protease.

Interestingly, a significant amount of Erv1 aggregated in the absence of Yme1. Erv1 is the sulfhydryl oxidase of the disulfide relay system and itself a substrate of Mia40 (Hell, 2008; Kallergi *et al.*, 2012). Until now, it was assumed that the introduction of disulfide bonds into the substrates of the Mia40-Erv1 disulfide relay system is a necessary and sufficient step to trigger folding of the entire, typically small substrate proteins. Erv1 is, however, significantly larger than most substrates of the disulfide relay system. The disulfide bond that is introduced by Mia40 is located in a very distal part of the polypeptide chain. It is conceivable that in the case of Erv1, the introduction of the disulfide bond is not sufficient for folding of the entire protein. Yme1 may thus be required to fold the residual part of the polypeptide chain of chain of Erv1.

In conclusion, the Yme1 substrates identified in the SILAC screen are certainly not uniform in terms of their import pathways, localization, size, structure and function. This points to a broad influence of Yme1 on diverse subsets of intermembrane space proteins. With this in mind, one can imagine a highly interrelated and interconnected network of folding and translocation systems, which were previously considered unrelated. Regarding the diversity of mitochondrial precursor proteins, an translocation and folding system that is fine-tuned and adapted to the needs of each individual substrate is not unlikely. Besides, it is almost normal that proposed molecular mechanisms are revealed to be more complex than previously expected. It will still be a long time, with intensive research in this field, before all interconnections between the diverse translocation and folding pathways are revealed and understood.

I observed that the endogenous levels of most of the newly identified Yme1 substrates are unaffected in the absence of Yme1, suggesting that Yme1 is not involved in the degradation of these substrates. This indicates that Yme1 is able to uncouple the proteolytic activity and the chaperone-like activity under certain circumstances. Chaperones and proteases closely interact in the maintenance of protein homeostasis, and there are several examples in the protein quality control systems where both functions are present on the same polypeptide chain. FtsH and LON protease both have this double function, whereas ClpP protease has to assemble with ClpX, a separate partner chaperone, to form the functional AAA protease (Moliere and Turgay, 2009). However, both components, ClpX and ClpP alone can also act independently under certain circumstances. It remains to be elucidated if all AAA proteases can switch from the degradation to the chaperoning mode and which factors determine this transition (Sauer and Baker, 2011). It is possible that Yme1 acts as chaperone in a first step, assisting the folding of proteins. After unsuccessful folding attempts, Yme1 could switch to the proteolytic mode and degrade the non-folded substrates in order to prevent deleterious effects on the rest of the cell. It is even conceivable that Yme1 acts as remodeler or disaggregase, converting aggregated or misfolded proteins back to a folding-competent state. It is interesting that for the *m*-AAA protease, such an isolated ATPase activity has already been reported. The *m*-AAA protease mediates membrane dislocation of the precursor of Ccp1 as a prerequisite for cleavage of Ccp1 by the rhomboid protease Pcp1 (Esser *et al.*, 2002). In this process, the proteolytic activity of *m*-AAA protease is dispensable (Tatsuta and Langer, 2007).

Gut2, one of the newly identified endogenous substrates of Yme1, was specifically co-isolated with Yme1 in the presence of ADP. In the presence of ATP, however, no co-isolation was detected. This reminds one of typical substrate-chaperone interactions. In the presence of the ATP, the interaction is too short-lived to be detected. In the presence of ADP, the substrate binds to the chaperone but does not get processed. Therefore, the interaction lasts longer and is thus easier to trace.

Besides Yme1, Mgr1 and Mgr3 were specifically co-isolated with IMS-DHFR^{WT} by Ni-NTA pulldown. Mgr1 and Mgr3 are anchored in the mitochondrial inner membrane and form a supermolecular complex with Yme1. Both proteins have been suggested to function as adaptors for the proteolytic substrates of Yme1 (Dunn *et al.*, 2008). For future experiments, it will be important to analyze *MGR1*, *MGR3* or double deletion strains in detail. The steady state levels and the folding state of the model substrate IMS-DHFR^{WT} and of the newly identified endogenous Yme1 substrates in this background should be determined. This could answer the question of whether the adaptor proteins are needed only for efficient processing of proteolytic substrates or also for the substrates of the chaperone-like activity.

4.4 Folding in the mitochondrial intermembrane space - unconventional

pathways

Members of the classical chaperone families have not been detected in the mitochondrial intermembrane space so far. However, the presence of atypical chaperones cannot be excluded. The situation is further complicated by the fact that some known chaperones, such as trigger factor in bacteria or the group of small heat shock proteins, operate in an ATP-independent mode (Ferbitz *et al.*, 2004; Haslbeck *et al.*, 2005). This makes their identification even more difficult.

Additionally, a dual localization of a number of origininally "only cytosolic" or "only mitochondrial" proteins seems to be more abundant than expected. In that sense, one quarter of the mitochondrial proteome has a dual localization (Dinur-Mills *et al.*, 2008; Yogev and Pines, 2011). Interestingly, the dual localized proteins, designated echoforms, often exert completely different functions in the different compartments (Ben-Menachem *et al.*, 2011; Yogev *et al.*, 2011). This underlines the possibility of the presence of classical cytosolic chaperones in the intermembrane space of mitochondria that have simply not been detected yet. And moreover, non-chaperone proteins of the cytosol or other cellular compartments may have "echoforms" in mitochondria that function as chaperones in this compartment.

A recent study in fungi on peroxisomal proteostasis suggested that Pln, a representative of the LON protease family, exerts a chaperone-like function (Bartoszewska *et al.*, 2012). Pln, Yme1 and other AAA proteases could possibly function in several different modes, ranging from chaperone to chaperone-protease-coupled, or protease activity. This finding underlines the possibility that other "well-known" proteins may also have in fact many more functional modes than expected and investigated until today.

Furthermore, some components of the mitochondrial translocation systems have recently been proposed to be involved in the folding of their substrates. It was shown that TOM complex can accommodate and transport a precursor protein containing two helices. This raises the possibility that folding could already start inside the TOM complex (Ahting et al., 1999). Moreover, it has been shown that TOM binds unfolded polypeptides and by this prevents their aggregation (Esaki et al., 2003). Due to the limited space inside the translocation channel, proteins will probably not fold completely with the help of the TOM complex. However, it could provide the initiating folding step and further folding of the protein would be taken over by downstream chaperones.

Notably, the small TIM proteins have been suggested to exert a chaperone-like activity in the mitochondrial intermembrane space. Small TIM complexes shield the hydrophobic residues of incoming preproteins and escort these proteins through the intermembrane space to the TOB translocase in the outer membrane, or to the TIM22 translocase in the inner membrane (Neupert and Herrmann, 2007; Chacinska et al., 2009; Endo and Yamano, 2009). It is conceivable that folding of some substrate proteins already occurs during transport through the intermembrane space. And indeed, recent evidence confirms that the precursor of Tim23 can partly fold upon interaction with Tim8-Tim13 and Tim9-Tim10 complex (Davis et al., 2007). Similarly, the intermembrane space domains of Tim50 and Tim23 could serve as folding platforms for substrates of the TIM23 complex. In this context, it was recently shown that the intermembrane space domain of Tim23 is intrinsically disordered (Gevorkyan-Airapetov et al., 2009; de la Cruz et al., 2010). Recently, such domains were suggested to exert chaperone-like functions (Dyson and Wright, 2005).

Similar to the TOM complex, Yme1 could provide a folding platform triggering the initial folding steps of translocating preproteins already during import. Subsequently, it could pass them on to downstream chaperones. The initial folding steps could even have a function in promotion of the ongoing import and prevention from retrograde transport of parts of the polypeptide. In this context, it is interesting to see that Yme1 seems to be involved in the import of human polynucleotide phosphorylase (PNPase) into yeast mitochondria (Rainey et al., 2006). However, PNPase is an ectopically expressed protein and it remains to be elucidated if Yme1 is also involved in the import of endogenous proteins and whether this applies for all or only a special subset of imported polypeptides.

In conclusion, it seems evident that the folding systems of the intermembrane space are rather "unconventional". They seem to differ significantly from all other known chaperone systems present in the cytosol of bacteria and eukaryotes or in the mitochondrial matrix. Moreover, it is obvious that several diverse folding pathways exist. These pathways are interconnected and cooperate with each other according to the substrate to be handled.

4.5 The human i-AAA protease

A recent study addressed the characterization of YME1L, the human ortholog of Yme1 (Coppola *et al.*, 2000; Shah *et al.*, 2000; Stiburek *et al.*, 2012). Stiburek *et al.* showed that YME1L resembles its yeast counterpart in many aspects. Similar to Yme1, YME1L is anchored to the mitochondrial inner membrane and forms complexes of high molecular weight. Furthermore, stable knockdown of *YME1L* in mammalian cells by an elegant RNA interference approach revealed a pleiotropic phenotype that overlaps in many aspects with the phenotype of the deletion of *YME1* in yeast. The *YME1L* knockout cells showed a respiratory deficiency that was attributed to an impaired activity of complex I of the respiratory chain. YME1L is one of the proteases that mediate the processing of the dynamin-like GTPase OPA1. The ratio of long to short isoforms of OPA1 determines mitochondrial dynamics (fusion and fission) and cristae morphology (Griparic *et al.*, 2007; Song *et al.*, 2007). The strong morphological phenotype of *YME1L* knockdown cells with its markedly altered mitochondrial network and a largely disorganized cristae morphology is probably attributed to an altered ratio of long to short OPA1 isoforms. Furthermore, *YME1L* knockout cells showed a significant decrease in cell proliferation that is, at least in part, a result of an increased rate of apoptosis. The increased susceptibility of *YME1L* knockout cells to programmed cell death seems to result from an impaired prevention of the accumulation of oxidatively damaged mitochondrial membrane proteins.

4.6 Protein quality control in the light of neurodegenerative diseases

For future research, it will be of great interest to characterize the folding system of the mitochondrial intermembrane space in the context of neurodegenerative diseases and ageing. During the last years, scientists revealed extensive connections between defects in the protein quality control systems and such neurological disorders as hereditary spastic paraplegia (HSP) and spinocerebellar ataxia (SCA) (Koppen and Langer, 2007; Tatsuta and Langer, 2008). In both cases, mutations of the human *m*-AAA protease, the matrix-facing counterpart of YME1L, are the cause for pathogenesis. Similarly, the prominent neurological disorders Parkinson's disease, Alzheimer's disease and Huntington's disease, which share the occurrance of amyloidogenic aggregates, are based on the failure of protein quality systems (Muchowski, 2002; Ross and Poirier, 2004; Muchowski and Wacker, 2005; Ross and Poirier, 2005; Morimoto, 2008). Again, pathogenesis is induced by the dysfunction of components of the protein quality control system. In Alzheimer's disease, the matrix protease PreP (Falkevall *et al.*, 2006; Rugarli and Langer, 2012) or components of the γ-secretase (Hedskog *et al.*, 2011; Pavlov *et al.*, 2011) are mutated. In Parkinson's disease, mutations have been indentified in a multitude of components of the protein quality control systems: Htr2A/OMI (Strauss *et al.*, 2005; Kawamoto *et al.*, 2008) PARL (Rugarli and Langer, 2012), Parkin (Kitada *et al.*, 1998) or PINK1 (Valente *et al.*, 2004).

The tissue-specific manifestation of these diseases is striking. It can be explained by some common characteristics of neurons, the cells that are usually most affected. Neurons have a very special shape with extremely long axons and a multitude of dendrites. Thus, they rely -more than other cells- on functional mitochondria that provide the enormous amounts of energy needed for neuronal signaling. As neurons are post-mitotic and terminally differentiated, they are strictly dependent on intact mitochondria. Mitochondria are the main suppliers of energy in the form of ATP. Therefore, they are strongly exposed to reactive oxygen species that are the by-products of oxidative phosphorylation (Murphy *et al.*, 2011). Reactive oxygen species cause damage of DNA and proteins. Mitochondrial DNA repair mechanisms and protein quality control systems can restore the integrity of these macromolecules. The capacity of the mitochondrial quality control system can even be expanded by the upregulation of mitochondrial chaperones and proteases. The increased transcription and translation of components of the mitochondrial quality control system is induced by the accumulation of misfolded or unfolded proteins. This control circuit is therefore referred to as mitochondrial unfolded protein response (Zhao *et al.*, 2002; Tatsuta, 2009; Haynes and Ron, 2010).

DISCUSSION

With increasing age, the capacity of the protein quality control systems seems to decrease dramatically and thus, misfolded protein species accumulate in a stepwise manner. This explains the late onset of many of these neurodegenerative diseases (Ross and Poirier, 2005; Hartl *et al.*, 2011). Protein aggregates are the characteristic feature of most neurodegenerative diseases. However, it is not yet clear if these aggregates are cytotoxic. Recent studies suggest that aggregates are indeed toxic because they sequester and block components of the protein quality control systems. Thus, the sequestered components of the protein quality control systems are unavailable in cases of cellular stress emergency (Brignull *et al.*, 2007; Chen *et al.*, 2011; Olzscha *et al.*, 2011).

With regard to diagnosis and treatment of the growing list of neurodegenerative diseases, it will be of particular interest to unravel the underlying mechanisms on the molecular level in detail. It is conceivable that mutations of YME1L in humans play a crucial role in the development of diverse neurodegenerative diseases, similar to the role of *m*-AAA protease in hereditary spastic paraplegia and spinocerebellar ataxia. To this end, further endogenous substrates of Yme1/YME1L have to be identified and analyzed. It will be a great challenge to find common principles on the molecular level and such principles could constitute the starting point for the development of adequate treatments. It is also possible, however, that each patient develops the disease on the basis of a largely unique and individual set of causes. In this case, it will be necessary to record the entire pattern of changes on the molecular level in order to develop an individually tailored treatment. This approach is already used in cancer treatment and is known as "personalized medicine" (Bianchi, 2012; de Cuba *et al.*, 2012).

In the future, it will be particularly important to promote interdisciplinary studies, combining basic research, genetics and clinical research. Only an interdisciplinary approach can reveal the correlations between molecular events and clinical symptoms and finally find causal therapy.

5. SUMMARY

The vast majority of mitochondrial proteins are encoded in the nucleus and synthesized as precursor proteins on cytosolic ribosomes. After translation, these precursor proteins are imported in a largely, if not completely, unfolded state into one of the four mitochondrial subcompartments, the outer membrane, the intermembrane space, the inner membrane or the matrix.

Once the precursor proteins reach their compartment of destination, they can fold into the functionally active three-dimensional native structure. Therefore, internal mitochondrial folding systems are needed in each subcompartment to assist folding of these precursor proteins upon import. Members of several "classical" chaperone families are present in the mitochondrial matrix and have been shown to support import and folding of newly imported polypeptides. However, folding of proteins in the mitochondrial intermembrane space is only poorly understood. Recently, a disulfide relay system in the intermembrane space that mediates import and folding was described, but this system is limited to proteins that form disulfide bonds. For the majority of intermembrane proteins, folding helpers that promote folding have not yet been discovered.

In order to identify general folding helpers of the intermembrane space, the well studied model substrate mouse dihydrofolate reductase (DHFR) was targeted to the mitochondrial intermembrane space of *S. cerevisiae* and its folding analyzed. DHFR assumes its mature fold in the intermembrane space and heat shock induces DHFR aggregation. Interestingly, aggregation is counteracted by an ATP-dependent process. The *i*-AAA protease Yme1 that is anchored in the inner mitochondrial membrane and exposes its functional domains to the intermembrane space was able to prevent the aggregation of DHFR.

A number of proteins of diverse structural and functional classes were found in the aggregate fractions of mitochondria lacking Yme1. Amongst them were factors that are involved in the establishment and maintenance of the mitochondrial ultrastructure, lipid metabolism, protein translocation and respiratory growth.

SUMMARY

Considering the diversity of the proteins affected in the absence of Yme1 and their function in mitochondria, the pleiotropic effects of the deletion of Yme1 can be readily explained. The findings of the present *in vivo* study confirm previous hints to a chaperone-like function of Yme1 resulting from *in vitro* experiments. Yme1 thus has a dual role as protease and as chaperone and occupies a key position in the protein quality control system of the mitochondrial intermembrane space.

6. LITERATURE

Ahting, U., Thun, C., Hegerl, R., Typke, D., Nargang, F.E., Neupert, W., and Nussberger, S. (1999). The TOM core complex: the general protein import pore of the outer membrane of mitochondria. J. Cell Biol. *147*, 959-968.
Anfinsen, C.B. (1973). Principles that govern the folding of protein chains. Science *181*, 223-230.
Arlt, H., Tauer, R., Feldmann, H., Neupert, W., and Langer, T. (1996). The YTA10-12 complex, an AAA protease with chaperone-like activity in the inner membrane of mitochondria. Cell *85*, 875-885.
Arretz, M., Schneider, H., Wienhues, U., and Neupert, W. (1991). Processing of mitochondrial precursor proteins. Biomed Biochim Acta *50*, 403-412.
Augustin, S., Nolden, M., Muller, S., Hardt, O., Arnold, I., and Langer, T. (2005). Characterization of peptides released from mitochondria: evidence for constant proteolysis and peptide efflux. J Biol Chem *280*, 2691-2699.
Baker, B.M., and Haynes, C.M. (2011). Mitochondrial protein quality control during biogenesis and aging. Trends Biochem Sci *36*, 254-261.
Baker, M.J., Tatsuta, T., and Langer, T. (2011). Quality control of mitochondrial proteostasis. Cold Spring Harb Perspect Biol *3*.
Baker, T.A., and Sauer, R.T. (2012). ClpXP, an ATP-powered unfolding and protein-degradation machine. Biochim Biophys Acta *1823*, 15-28.
Balch, W.E., Morimoto, R.I., Dillin, A., and Kelly, J.W. (2008). Adapting proteostasis for disease intervention. Science *319*, 916-919.
Banci, L., Bertini, I., Cefaro, C., Cenacchi, L., Ciofi-Baffoni, S., Felli, I.C., Gallo, A., Gonnelli, L., Luchinat, E., Sideris, D., and Tokatlidis, K. (2010). Molecular chaperone function of Mia40 triggers consecutive induced folding steps of the substrate in mitochondrial protein import. Proc Natl Acad Sci U S A *107*, 20190-20195.
Bartlett, A.I., and Radford, S.E. (2009). An expanding arsenal of experimental methods yields an explosion of insights into protein folding mechanisms. Nat Struct Mol Biol *16*, 582-588.
Bartoszewska, M., Williams, C., Kikhney, A., Opalinski, L., van Roermund, C.W., de Boer, R., Veenhuis, M., and van der Klei, I.J. (2012). Peroxisomal proteostasis involves a lon family protein that functions as protease and chaperone. J Biol Chem *287*, 27380-27395.
Beasley, E.M., Muller, S., and Schatz, G. (1993). The signal that sorts yeast cytochrome b2 to the mitochondrial intermembrane space contains three distinct functional regions. EMBO J. *12*, 2303-2311.
Becker, T., Bottinger, L., and Pfanner, N. (2012). Mitochondrial protein import: from transport pathways to an integrated network. Trends Biochem Sci *37*, 85-91.
Becker, T., Pfannschmidt, S., Guiard, B., Stojanovski, D., Milenkovic, D., Kutik, S., Pfanner, N., Meisinger, C., and Wiedemann, N. (2008). Biogenesis of the mitochondrial TOM complex: Mim1 promotes insertion and assembly of signal-anchored receptors. J Biol Chem *283*, 120-127.

Ben-Menachem, R., Tal, M., Shadur, T., and Pines, O. (2011). A third of the yeast mitochondrial proteome is dual localized: a question of evolution. Proteomics *11*, 4468-4476.
Ben-Zvi, A.P., and Goloubinoff, P. (2001). Review: mechanisms of disaggregation and refolding of stable protein aggregates by molecular chaperones. J Struct Biol *135*, 84-93.
Berger, K.H., and Yaffe, M.P. (1998). Prohibitin family members interact genetically with mitochondrial inheritance components in Saccharomyces cerevisiae. Mol Cell Biol *18*, 4043-4052.
Beyer, A. (1997). Sequence analysis of the AAA protein family. Protein Sci *6*, 2043-2058.
Bianchi, D.W. (2012). From prenatal genomic diagnosis to fetal personalized medicine: progress and challenges. Nat Med *18*, 1041-1051.
Bieniossek, C., Schalch, T., Bumann, M., Meister, M., Meier, R., and Baumann, U. (2006). The molecular architecture of the metalloprotease FtsH. Proc Natl Acad Sci U S A *103*, 3066-3071.
Bieschke, J., Cohen, E., Murray, A., Dillin, A., and Kelly, J.W. (2009). A kinetic assessment of the C. elegans amyloid disaggregation activity enables uncoupling of disassembly and proteolysis. Protein Sci *18*, 2231-2241.
Birnboim, H., and Doly, J. (1979). A rapid alkaline extraction procedure for screening recombinant plasmid DNA. NAR *11*, 4077-4092.
Bohnert, M., Rehling, P., Guiard, B., Herrmann, J.M., Pfanner, N., and van der Laan, M. (2010). Cooperation of stop-transfer and conservative sorting mechanisms in mitochondrial protein transport. Curr Biol *20*, 1227-1232.
Bolliger, L., Deloche, O., Glick, B.S., Georgopoulos, C., Jeno, P., Kronidou, N., Horst, M., Morishima, N., and Schatz, G. (1994). A mitochondrial homolog of bacterial GrpE interacts with mitochondrial hsp70 and is essential for viability. EMBO J. *13*, 1998-2006.
Bota, D.A., Ngo, J.K., and Davies, K.J. (2005). Downregulation of the human Lon protease impairs mitochondrial structure and function and causes cell death. Free Radic Biol Med *38*, 665-677.
Bota, D.A., Van Remmen, H., and Davies, K.J. (2002). Modulation of Lon protease activity and aconitase turnover during aging and oxidative stress. FEBS Lett *532*, 103-106.
Botos, I., Melnikov, E.E., Cherry, S., Tropea, J.E., Khalatova, A.G., Rasulova, F., Dauter, Z., Maurizi, M.R., Rotanova, T.V., Wlodawer, A., and Gustchina, A. (2004). The catalytic domain of Escherichia coli Lon protease has a unique fold and a Ser-Lys dyad in the active site. J Biol Chem *279*, 8140-8148.
Brignull, H.R., Morley, J.F., and Morimoto, R.I. (2007). The stress of misfolded proteins: C. elegans models for neurodegenerative disease and aging. Adv Exp Med Biol *594*, 167-189.
Brinker, A., Pfeifer, G., Kerner, M.J., Naylor, D.J., Hartl, F.U., and Hayer-Hartl, M. (2001). Dual function of protein confinement in chaperonin-assisted protein folding. Cell *107*, 223-233.
Buchberger, A., Bukau, B., and Sommer, T. (2010). Protein quality control in the cytosol and the endoplasmic reticulum: brothers in arms. Mol Cell *40*, 238-252.
Bukau, B., and Horwich, A.L. (1998). The Hsp70 and Hsp60 chaperone machines. Cell *92*, 351-366.
Bukau, B., Weissman, J., and Horwich, A. (2006). Molecular chaperones and protein quality control. Cell *125*, 443-451.

Burri, L., Vascotto, K., Fredersdorf, S., Tiedt, R., Hall, M.N., and Lithgow, T. (2004). Zim17, a novel zinc finger protein essential for protein import into mitochondria. J Biol Chem *279*, 50243-50249.

Campbell, C.L., Tanaka, N., White, K.H., and Thorsness, P.E. (1994). Mitochondrial morphological and functional defects in yeast caused by yme1 are suppressed by mutation of a 26S protease subunit homologue. Mol Biol Cell *5*, 899-905.

Campbell, C.L., and Thorsness, P.E. (1998). Escape of mitochondrial DNA to the nucleus in yme1 yeast is mediated by vacuolar-dependent turnover of abnormal mitochondrial compartments. J Cell Sci *111 (Pt 16)*, 2455-2464.

Cha, S.S., An, Y.J., Lee, C.R., Lee, H.S., Kim, Y.G., Kim, S.J., Kwon, K.K., De Donatis, G.M., Lee, J.H., Maurizi, M.R., and Kang, S.G. (2010). Crystal structure of Lon protease: molecular architecture of gated entry to a sequestered degradation chamber. EMBO J *29*, 3520-3530.

Chacinska, A., Koehler, C.M., Milenkovic, D., Lithgow, T., and Pfanner, N. (2009). Importing mitochondrial proteins: machineries and mechanisms. Cell *138*, 628-644.

Chakraborty, K., Chatila, M., Sinha, J., Shi, Q., Poschner, B.C., Sikor, M., Jiang, G., Lamb, D.C., Hartl, F.U., and Hayer-Hartl, M. (2010). Chaperonin-catalyzed rescue of kinetically trapped states in protein folding. Cell *142*, 112-122.

Chen, B., Retzlaff, M., Roos, T., and Frydman, J. (2011). Cellular strategies of protein quality control. Cold Spring Harb Perspect Biol *3*, a004374.

Chiba, S., Ito, K., and Akiyama, Y. (2006). The Escherichia coli plasma membrane contains two PHB (prohibitin homology) domain protein complexes of opposite orientations. Mol Microbiol *60*, 448-457.

Chiti, F., and Dobson, C.M. (2006). Protein misfolding, functional amyloid, and human disease. Annu Rev Biochem *75*, 333-366.

Cooper, C.E., Nicholls, P., and Freedman, J.A. (1991). Cytochrome c oxidase: structure, function, and membrane topology of the polypeptide subunits. Biochem Cell Biol *69*, 586-607.

Coppola, M., Pizzigoni, A., Banfi, S., Bassi, M.T., Casari, G., and Incerti, B. (2000). Identification and characterization of YME1L1, a novel paraplegin-related gene. Genomics *66*, 48-54.

D'Silva, P.D., Schilke, B., Walter, W., Andrew, A., and Craig, E.A. (2003). J protein cochaperone of the mitochondrial inner membrane required for protein import into the mitochondrial matrix. Proc Natl Acad Sci U S A *100*, 13839-13844.

Daum, G., Gasser, S.M., and Schatz, G. (1982). Import of proteins into mitochondria. Energy-dependent, two-step processing of the intermembrane space enzyme cytochrome b2 by isolated yeast mitochondria. J. Biol. Chem. *257*, 13075-13080.

Davis, A.J., Alder, N.N., Jensen, R.E., and Johnson, A.E. (2007). The Tim9p/10p and Tim8p/13p complexes bind to specific sites on Tim23p during mitochondrial protein import. Mol Biol Cell *18*, 475-486.

de Cuba, E.M., Kwakman, R., van Egmond, M., Bosch, L.J., Bonjer, H.J., Meijer, G.A., and te Velde, E.A. (2012). Understanding molecular mechanisms in peritoneal dissemination of colorectal cancer : future possibilities for personalised treatment by use of biomarkers. Virchows Arch *461*, 231-243.

de la Cruz, L., Bajaj, R., Becker, S., and Zweckstetter, M. (2010). The intermembrane space domain of Tim23 is intrinsically disordered with a distinct binding region for presequences. Protein Sci *19*, 2045-2054.

Deponte, M., and Hell, K. (2009). Disulphide bond formation in the intermembrane space of mitochondria. J Biochem *146*, 599-608.
Dimmer, K.S., Papic, D., Schumann, B., Sperl, D., Krumpe, K., Walther, D.M., and Rapaport, D. (2012). A crucial role for Mim2 in the biogenesis of mitochondrial outer membrane proteins. J Cell Sci *125*, 3464-3473.
Dinur-Mills, M., Tal, M., and Pines, O. (2008). Dual targeted mitochondrial proteins are characterized by lower MTS parameters and total net charge. PLoS One *3*, e2161.
Dobson, C.M., and Karplus, M. (1999). The fundamentals of protein folding: bringing together theory and experiment. Curr Opin Struct Biol *9*, 92-101.
Dollins, D.E., Warren, J.J., Immormino, R.M., and Gewirth, D.T. (2007). Structures of GRP94-nucleotide complexes reveal mechanistic differences between the hsp90 chaperones. Mol Cell *28*, 41-56.
Dougan, D.A., Reid, B.G., Horwich, A.L., and Bukau, B. (2002). ClpS, a substrate modulator of the ClpAP machine. Mol Cell *9*, 673-683.
Douglas, N.R., Reissmann, S., Zhang, J., Chen, B., Jakana, J., Kumar, R., Chiu, W., and Frydman, J. (2011). Dual action of ATP hydrolysis couples lid closure to substrate release into the group II chaperonin chamber. Cell *144*, 240-252.
Dudkina, N.V., Kouril, R., Peters, K., Braun, H.P., and Boekema, E.J. (2010). Structure and function of mitochondrial supercomplexes. Biochim Biophys Acta *1797*, 664-670.
Dumont, M.E., Cardillo, T.S., Hayes, M.K., and Sherman, F. (1991). Role of cytochrome c heme lyase in mitochondrial import and accumulation of cytochrome c in Saccharomyces cerevisiae. Mol. Cell. Biol. *11*, 5487-5496.
Dunn, C.D., Lee, M.S., Spencer, F.A., and Jensen, R.E. (2006). A genomewide screen for petite-negative yeast strains yields a new subunit of the i-AAA protease complex. Mol Biol Cell *17*, 213-226.
Dunn, C.D., Tamura, Y., Sesaki, H., and Jensen, R.E. (2008). Mgr3p and Mgr1p are adaptors for the mitochondrial i-AAA protease complex. Mol Biol Cell *19*, 5387-5397.
Dyson, H.J., and Wright, P.E. (2005). Intrinsically unstructured proteins and their functions. Nat Rev Mol Cell Biol *6*, 197-208.
Eilers, M., and Schatz, G. (1986). Binding of a specific ligand inhibits import of a purified precursor protein into mitochondria. Nature *322*, 228-232.
Elliott, L.E., Saracco, S.A., and Fox, T.D. (2011). Multiple Roles of the Cox20 Chaperone in Assembly of Saccharomyces cerevisiae Cytochrome c Oxidase. Genetics.
Ellis, R.J., and Minton, A.P. (2006). Protein aggregation in crowded environments. Biol Chem *387*, 485-497.
Elsasser, S., and Finley, D. (2005). Delivery of ubiquitinated substrates to protein-unfolding machines. Nat Cell Biol *7*, 742-749.
Endo, T., and Yamano, K. (2009). Multiple pathways for mitochondrial protein traffic. Biol Chem *390*, 723-730.
Erbse, A., Mayer, M.P., and Bukau, B. (2004). Mechanism of substrate recognition by Hsp70 chaperones. Biochem Soc Trans *32*, 617-621.
Erbse, A., Schmidt, R., Bornemann, T., Schneider-Mergener, J., Mogk, A., Zahn, R., Dougan, D.A., and Bukau, B. (2006). ClpS is an essential component of the N-end rule pathway in Escherichia coli. Nature *439*, 753-756.

Esaki, M., Kanamori, T., Nishikawa, S.I., Shin, I., Schultz, P.G., and Endo, T. (2003). Tom40 protein import channel binds to non-native proteins and prevents their aggregation. Nat Struct Biol.
Esser, K., Tursun, B., Ingenhoven, M., Michaelis, G., and Pratje, E. (2002). A novel two-step mechanism for removal of a mitochondrial signal sequence involves the mAAA complex and the putative rhomboid protease Pcp1. J Mol Biol 323, 835-843.
Falkevall, A., Alikhani, N., Bhushan, S., Pavlov, P.F., Busch, K., Johnson, K.A., Eneqvist, T., Tjernberg, L., Ankarcrona, M., and Glaser, E. (2006). Degradation of the amyloid beta-protein by the novel mitochondrial peptidasome, PreP. J Biol Chem 281, 29096-29104.
Felts, S.J., Owen, B.A., Nguyen, P., Trepel, J., Donner, D.B., and Toft, D.O. (2000). The hsp90-related protein TRAP1 is a mitochondrial protein with distinct functional properties. J Biol Chem 275, 3305-3312.
Ferbitz, L., Maier, T., Patzelt, H., Bukau, B., Deuerling, E., and Ban, N. (2004). Trigger factor in complex with the ribosome forms a molecular cradle for nascent proteins. Nature 431, 590-596.
Field, L.S., Furukawa, Y., O'Halloran, T.V., and Culotta, V.C. (2003). Factors controlling the uptake of yeast copper/zinc superoxide dismutase into mitochondria. J Biol Chem 278, 28052-28059.
Fiumera, H.L., Dunham, M.J., Saracco, S.A., Butler, C.A., Kelly, J.A., and Fox, T.D. (2009). Translocation and assembly of mitochondrially coded Saccharomyces cerevisiae cytochrome c oxidase subunit Cox2 by Oxa1 and Yme1 in the absence of Cox18. Genetics 182, 519-528.
Flynn, J.M., Levchenko, I., Sauer, R.T., and Baker, T.A. (2004). Modulating substrate choice: the SspB adaptor delivers a regulator of the extracytoplasmic-stress response to the AAA+ protease ClpXP for degradation. Genes Dev 18, 2292-2301.
Frey, S., Leskovar, A., Reinstein, J., and Buchner, J. (2007). The ATPase cycle of the endoplasmic chaperone Grp94. J Biol Chem 282, 35612-35620.
Frickey, T., and Lupas, A.N. (2004). Phylogenetic analysis of AAA proteins. J Struct Biol 146, 2-10.
Frydman, J. (2001). Folding of newly translated proteins in vivo: the role of molecular chaperones. Annu Rev Biochem 70, 603-647.
Gangjee, A., Li, W., Lin, L., Zeng, Y., Ihnat, M., Warnke, L.A., Green, D.W., Cody, V., Pace, J., and Queener, S.F. (2009). Design, synthesis, and X-ray crystal structures of 2,4-diaminofuro[2,3-d]pyrimidines as multireceptor tyrosine kinase and dihydrofolate reductase inhibitors. Bioorg Med Chem 17, 7324-7336.
Gaume, B., Klaus, C., Ungermann, C., Guiard, B., Neupert, W., and Brunner, M. (1998). Unfolding of preproteins upon import into mitochondria. EMBO J. 17, 6497-6507.
Geissler, A., Krimmer, T., Bomer, U., Guiard, B., Rassow, J., and Pfanner, N. (2000). Membrane potential-driven protein import into mitochondria. The sorting sequence of cytochrome b(2) modulates the deltapsi-dependence of translocation of the matrix-targeting sequence. Mol. Biol. Cell 11, 3977-3991.
Gerdes, F., Tatsuta, T., and Langer, T. (2012). Mitochondrial AAA proteases - Towards a molecular understanding of membrane-bound proteolytic machines. Biochim Biophys Acta 1823, 49-55.
Gevorkyan-Airapetov, L., Zohary, K., Popov-Celeketic, D., Mapa, K., Hell, K., Neupert, W., Azem, A., and Mokranjac, D. (2009). Interaction of Tim23 with

Tim50 Is essential for protein translocation by the mitochondrial TIM23 complex. J Biol Chem *284*, 4865-4872.
Goldberg, A.L., and Waxman, L. (1985). The role of ATP hydrolysis in the breakdown of proteins and peptides by protease La from Escherichia coli. J Biol Chem *260*, 12029-12034.
Graef, M., Seewald, G., and Langer, T. (2007). Substrate recognition by AAA+ ATPases: distinct substrate binding modes in ATP-dependent protease Yme1 of the mitochondrial intermembrane space. Mol Cell Biol *27*, 2476-2485.
Gray, M.W. (1999). Evolution of organellar genomes. Curr Opin Genet Dev *9*, 678-687.
Griparic, L., Kanazawa, T., and van der Bliek, A.M. (2007). Regulation of the mitochondrial dynamin-like protein Opa1 by proteolytic cleavage. J Cell Biol *178*, 757-764.
Gruhler, A., Ono, H., Guiard, B., Neupert, W., and Stuart, R.A. (1995). A novel intermediate on the import pathway of cytochrome b2 into mitochondria: evidence for conservative sorting. EMBO J. *14*, 1349-1359.
Gur, E., and Sauer, R.T. (2008). Recognition of misfolded proteins by Lon, a AAA(+) protease. Genes Dev *22*, 2267-2277.
Hanson, P.I., and Whiteheart, S.W. (2005). AAA+ proteins: have engine, will work. Nat Rev Mol Cell Biol *6*, 519-529.
Harel-Bronstein, M., Dibrov, P., Olami, Y., Pinner, E., Schuldiner, S., and Padan, E. (1995). MH1, a second-site revertant of an Escherichia coli mutant lacking Na+/H+ antiporters (delta nhaA delta nhaB), regains Na+ resistance and a capacity to excrete Na+ in a delta microH(+)-independent fashion. J Biol Chem *270*, 3816-3822.
Harner, M., Korner, C., Walther, D., Mokranjac, D., Kaesmacher, J., Welsch, U., Griffith, J., Mann, M., Reggiori, F., and Neupert, W. (2011). The mitochondrial contact site complex, a determinant of mitochondrial architecture. EMBO J *30*, 4356-4370.
Hartl, F.U. (1995). Principles of chaperone-mediated protein folding. Philos Trans R Soc Lond [Biol] *348*, 107-112.
Hartl, F.U., Bracher, A., and Hayer-Hartl, M. (2011). Molecular chaperones in protein folding and proteostasis. Nature *475*, 324-332.
Hartl, F.U., and Hayer-Hartl, M. (2009). Converging concepts of protein folding in vitro and in vivo. Nat Struct Mol Biol *16*, 574-581.
Haslbeck, M., Franzmann, T., Weinfurtner, D., and Buchner, J. (2005). Some like it hot: the structure and function of small heat-shock proteins. Nat Struct Mol Biol *12*, 842-846.
Haynes, C.M., and Ron, D. (2010). The mitochondrial UPR - protecting organelle protein homeostasis. J Cell Sci *123*, 3849-3855.
Hedskog, L., Petersen, C.A., Svensson, A.I., Welander, H., Tjernberg, L.O., Karlstrom, H., and Ankarcrona, M. (2011). gamma-Secretase complexes containing caspase-cleaved presenilin-1 increase intracellular Abeta(42) /Abeta(40) ratio. J Cell Mol Med *15*, 2150-2163.
Hell, K. (2008). The Erv1-Mia40 disulfide relay system in the intermembrane space of mitochondria. Biochim Biophys Acta *1783*, 601-609.
Herrmann, J.M., and Hell, K. (2005). Chopped, trapped or tacked--protein translocation into the IMS of mitochondria. Trends Biochem Sci *30*, 205-211.
Herrmann, J.M., and Neupert, W. (2000). What fuels polypeptide translocation? An energetical view on mitochondrial protein sorting. Biochim Biophys Acta *1459*, 331-338.

Herrmann, J.M., and Neupert, W. (2003). Protein insertion into the inner membrane of mitochondria. IUBMB Life *55*, 219-225.
Herrmann, J.M., and Riemer, J. (2012). Mitochondrial disulfide relay: redox-regulated protein import into the intermembrane space. J Biol Chem *287*, 4426-4433.
Hessling, M., Richter, K., and Buchner, J. (2009). Dissection of the ATP-induced conformational cycle of the molecular chaperone Hsp90. Nat Struct Mol Biol *16*, 287-293.
Hetz, C. (2012). The unfolded protein response: controlling cell fate decisions under ER stress and beyond. Nat Rev Mol Cell Biol *13*, 89-102.
Heyrovska, N., Frydman, J., Hohfeld, J., and Hartl, F.U. (1998). Directionality of polypeptide transfer in the mitochondrial pathway of chaperone-mediated protein folding. Biol Chem *379*, 301-309.
Hoppins, S., Collins, S.R., Cassidy-Stone, A., Hummel, E., Devay, R.M., Lackner, L.L., Westermann, B., Schuldiner, M., Weissman, J.S., and Nunnari, J. (2011). A mitochondrial-focused genetic interaction map reveals a scaffold-like complex required for inner membrane organization in mitochondria. J Cell Biol *195*, 323-340.
Hoppins, S., Lackner, L., and Nunnari, J. (2007). The machines that divide and fuse mitochondria. Annu Rev Biochem *76*, 751-780.
Horst, M., Oppliger, W., Rospert, S., Schonfeld, H.J., Schatz, G., and Azem, A. (1997). Sequential action of two hsp70 complexes during protein import into mitochondria. EMBO J. *16*, 1842-1849.
Horvath, A., and Riezman, H. (1994). Rapid protein extraction from Saccharomyces cerevisiae. Yeast *10*, 1305-1310.
Horwich, A.L., Farr, G.W., and Fenton, W.A. (2006). GroEL-GroES-mediated protein folding. Chem Rev *106*, 1917-1930.
Horwich, A.L., and Fenton, W.A. (2009). Chaperonin-mediated protein folding: using a central cavity to kinetically assist polypeptide chain folding. Q Rev Biophys *42*, 83-116.
Husnjak, K., Elsasser, S., Zhang, N., Chen, X., Randles, L., Shi, Y., Hofmann, K., Walters, K.J., Finley, D., and Dikic, I. (2008). Proteasome subunit Rpn13 is a novel ubiquitin receptor. Nature *453*, 481-488.
Inadome, H., Noda, Y., Adachi, H., and Yoda, K. (2001). A novel protein, Mpm1, of the mitochondria of the yeast Saccharomyces cerevisiae. Biosci Biotechnol Biochem *65*, 2577-2580.
Iosefson, O., Sharon, S., Goloubinoff, P., and Azem, A. (2012). Reactivation of protein aggregates by mortalin and Tid1--the human mitochondrial Hsp70 chaperone system. Cell Stress Chaperones *17*, 57-66.
Ito, K., and Akiyama, Y. (2005). Cellular functions, mechanism of action, and regulation of FtsH protease. Annu Rev Microbiol *59*, 211-231.
Iyer, L.M., and Aravind, L. (2004). The emergence of catalytic and structural diversity within the beta-clip fold. Proteins *55*, 977-991.
Iyer, L.M., Leipe, D.D., Koonin, E.V., and Aravind, L. (2004). Evolutionary history and higher order classification of AAA+ ATPases. J Struct Biol *146*, 11-31.
Junker, J.P., Hell, K., Schlierf, M., Neupert, W., and Rief, M. (2005). Influence of substrate binding on the mechanical stability of mouse dihydrofolate reductase. Biophys J *89*, L46-48.
Kallergi, E., Andreadaki, M., Kritsiligkou, P., Katrakili, N., Pozidis, C., Tokatlidis, K., Banci, L., Bertini, I., Cefaro, C., Ciofi-Baffoni, S., Gajda, K., and Peruzzini, R. (2012). Targeting and Maturation of Erv1/ALR in the Mitochondrial Intermembrane Space. ACS Chem Biol.

Kampinga, H.H., and Craig, E.A. (2010). The HSP70 chaperone machinery: J proteins as drivers of functional specificity. Nat Rev Mol Cell Biol *11*, 579-592.
Kanamori, T., Nishikawa, S., Shin, I., Schultz, P.G., and Endo, T. (1997). Probing the environment along the protein import pathways in yeast mitochondria by site-specific photocrosslinking. Proc Natl Acad Sci U S A *94*, 485-490.
Kang, S.G., Dimitrova, M.N., Ortega, J., Ginsburg, A., and Maurizi, M.R. (2005). Human mitochondrial ClpP is a stable heptamer that assembles into a tetradecamer in the presence of ClpX. J Biol Chem *280*, 35424-35432.
Kang, S.G., Ortega, J., Singh, S.K., Wang, N., Huang, N.N., Steven, A.C., and Maurizi, M.R. (2002). Functional proteolytic complexes of the human mitochondrial ATP-dependent protease, hClpXP. J Biol Chem *277*, 21095-21102.
Karata, K., Inagawa, T., Wilkinson, A.J., Tatsuta, T., and Ogura, T. (1999). Dissecting the role of a conserved motif (the second region of homology) in the AAA family of ATPases. Site-directed mutagenesis of the ATP-dependent protease FtsH. J Biol Chem *274*, 26225-26232.
Karata, K., Verma, C.S., Wilkinson, A.J., and Ogura, T. (2001). Probing the mechanism of ATP hydrolysis and substrate translocation in the AAA protease FtsH by modelling and mutagenesis. Mol Microbiol *39*, 890-903.
Kawamoto, Y., Kobayashi, Y., Suzuki, Y., Inoue, H., Tomimoto, H., Akiguchi, I., Budka, H., Martins, L.M., Downward, J., and Takahashi, R. (2008). Accumulation of HtrA2/Omi in neuronal and glial inclusions in brains with alpha-synucleinopathies. J Neuropathol Exp Neurol *67*, 984-993.
Khanday, F.A., Saha, M., and Bhat, P.J. (2002). Molecular characterization of MRG19 of Saccharomyces cerevisiae. Implication in the regulation of galactose and nonfermentable carbon source utilization. Eur J Biochem *269*, 5840-5850.
Khyse-Anderson, J. (1984). Electroblotting of multiple gels: a simple apparatus without buffer tank for rapid transfer of proteins from polyacrylamid to nitrocellulose. J. Biochem. Biophys. Methods *10*, 203-207.
Kitada, T., Asakawa, S., Hattori, N., Matsumine, H., Yamamura, Y., Minoshima, S., Yokochi, M., Mizuno, Y., and Shimizu, N. (1998). Mutations in the parkin gene cause autosomal recessive juvenile parkinsonism. Nature *392*, 605-608.
Koehler, C.M. (2004a). New developments in mitochondrial assembly. Annual Review of Cell and Developmental Biology *20*, 309-335.
Koehler, C.M. (2004b). The small Tim proteins and the twin Cx(3)C motif. Trends Biochem Sci *29*, 1-4.
Koehler, C.M., and Tienson, H.L. (2009). Redox regulation of protein folding in the mitochondrial intermembrane space. Biochim Biophys Acta *1793*, 139-145.
Kominsky, D.J., Brownson, M.P., Updike, D.L., and Thorsness, P.E. (2002). Genetic and biochemical basis for viability of yeast lacking mitochondrial genomes. Genetics *162*, 1595-1604.
Koppen, M., and Langer, T. (2007). Protein degradation within mitochondria: versatile activities of AAA proteases and other peptidases. Crit Rev Biochem Mol Biol *42*, 221-242.
Korbel, D., Wurth, S., Kaser, M., and Langer, T. (2004). Membrane protein turnover by the m-AAA protease in mitochondria depends on the transmembrane domains of its subunits. EMBO Rep *5*, 698-703.
Kozjak, V., Wiedemann, N., Milenkovic, D., Lohaus, C., Meyer, H.E., Guiard, B., Meisinger, C., and Pfanner, N. (2003). An essential role of Sam50 in the protein sorting and assembly machinery of the mitochondrial outer membrane. J Biol Chem *278*, 48520-48523.

Kress, W., Mutschler, H., and Weber-Ban, E. (2009). Both ATPase domains of ClpA are critical for processing of stable protein structures. J Biol Chem *284*, 31441-31452.
Krzewska, J., Langer, T., and Liberek, K. (2001). Mitochondrial Hsp78, a member of the Clp/Hsp100 family in Saccharomyces cerevisiae, cooperates with Hsp70 in protein refolding. FEBS Lett *489*, 92-96.
Kunau, W.H., Beyer, A., Franken, T., Gotte, K., Marzioch, M., Saidowsky, J., Skaletz-Rorowski, A., and Wiebel, F.F. (1993). Two complementary approaches to study peroxisome biogenesis in Saccharomyces cerevisiae: forward and reversed genetics. Biochimie *75*, 209-224.
Kushnirov, V.V. (2000). Rapid and reliable protein extraction from yeast. Yeast *16*, 857-860.
Langer, T., Kaser, M., Klanner, C., and Leonhard, K. (2001). AAA proteases of mitochondria: quality control of membrane proteins and regulatory functions during mitochondrial biogenesis. Biochem Soc Trans *29*, 431-436.
Langklotz, S., Baumann, U., and Narberhaus, F. (2012). Structure and function of the bacterial AAA protease FtsH. Biochim Biophys Acta *1823*, 40-48.
Lee, S., Sowa, M.E., Choi, J.M., and Tsai, F.T. (2004). The ClpB/Hsp104 molecular chaperone-a protein disaggregating machine. J Struct Biol *146*, 99-105.
Leidhold, C., and Voos, W. (2007). Chaperones and proteases--guardians of protein integrity in eukaryotic organelles. Ann N Y Acad Sci *1113*, 72-86.
Leitner, A., Joachimiak, L.A., Bracher, A., Monkemeyer, L., Walzthoeni, T., Chen, B., Pechmann, S., Holmes, S., Cong, Y., Ma, B., et al. (2012). The molecular architecture of the eukaryotic chaperonin TRiC/CCT. Structure *20*, 814-825.
Lemaire, C., Hamel, P., Velours, J., and Dujardin, G. (2000). Absence of the mitochondrial AAA protease Yme1p restores F0-ATPase subunit accumulation in an oxa1 deletion mutant of Saccharomyces cerevisiae. J Biol Chem *275*, 23471-23475.
Leonhard, K., Guiard, B., Pellecchia, G., Tzagoloff, A., Neupert, W., and Langer, T. (2000). Membrane protein degradation by AAA proteases in mitochondria: extraction of substrates from either membrane surface. Mol Cell *5*, 629-638.
Leonhard, K., Herrmann, J.M., Stuart, R.A., Mannhaupt, G., Neupert, W., and Langer, T. (1996). AAA proteases with catalytic sites on opposite membrane surfaces comprise a proteolytic system for the ATP-dependent degradation of inner membrane proteins in mitochondria. EMBO J *15*, 4218-4229.
Leonhard, K., Stiegler, A., Neupert, W., and Langer, T. (1999). Chaperone-like activity of the AAA domain of the yeast Yme1 AAA protease. Nature *398*, 348-351.
Leskovar, A., Wegele, H., Werbeck, N.D., Buchner, J., and Reinstein, J. (2008). The ATPase cycle of the mitochondrial Hsp90 analog Trap1. J Biol Chem *283*, 11677-11688.
Levchenko, I., Seidel, M., Sauer, R.T., and Baker, T.A. (2000). A specificity-enhancing factor for the ClpXP degradation machine. Science *289*, 2354-2356.
Lies, M., and Maurizi, M.R. (2008). Turnover of endogenous SsrA-tagged proteins mediated by ATP-dependent proteases in Escherichia coli. J Biol Chem *283*, 22918-22929.
Lill, R., Hoffmann, B., Molik, S., Pierik, A.J., Rietzschel, N., Stehling, O., Uzarska, M.A., Webert, H., Wilbrecht, C., and Muhlenhoff, U. (2012). The role of mitochondria in cellular iron-sulfur protein biogenesis and iron metabolism. Biochim Biophys Acta *1823*, 1491-1508.

Liu, Q., Krzewska, J., Liberek, K., and Craig, E.A. (2001). Mitochondrial Hsp70 Ssc1: role in protein folding. J Biol Chem *276*, 6112-6118.
Lotz, G.P., Legleiter, J., Aron, R., Mitchell, E.J., Huang, S.Y., Ng, C., Glabe, C., Thompson, L.M., and Muchowski, P.J. (2010). Hsp70 and Hsp40 functionally interact with soluble mutant huntingtin oligomers in a classic ATP-dependent reaction cycle. J Biol Chem *285*, 38183-38193.
Lu, L., Roberts, G., Simon, K., Yu, J., and Hudson, A.P. (2003). Rsf1p, a protein required for respiratory growth of Saccharomyces cerevisiae. Curr Genet *43*, 263-272.
Lupas, A., Flanagan, J.M., Tamura, T., and Baumeister, W. (1997). Self-compartmentalizing proteases. Trends Biochem Sci *22*, 399-404.
Lupas, A.N., and Martin, J. (2002). AAA proteins. Curr Opin Struct Biol *12*, 746-753.
Luttik, M.A., Overkamp, K.M., Kotter, P., de Vries, S., van Dijken, J.P., and Pronk, J.T. (1998). The Saccharomyces cerevisiae NDE1 and NDE2 genes encode separate mitochondrial NADH dehydrogenases catalyzing the oxidation of cytosolic NADH. J Biol Chem *273*, 24529-24534.
Marom, M., Azem, A., and Mokranjac, D. (2011). Understanding the molecular mechanism of protein translocation across the mitochondrial inner membrane: still a long way to go. Biochim Biophys Acta *1808*, 990-1001.
Martin, S.J. (2010). Cell biology. Opening the cellular poison cabinet. Science *330*, 1330-1331.
Martinelli, P., La Mattina, V., Bernacchia, A., Magnoni, R., Cerri, F., Cox, G., Quattrini, A., Casari, G., and Rugarli, E.I. (2009). Genetic interaction between the m-AAA protease isoenzymes reveals novel roles in cerebellar degeneration. Hum Mol Genet *18*, 2001-2013.
Matouschek, A., Pfanner, N., and Voos, W. (2000). Protein unfolding by mitochondria. The Hsp70 import motor. EMBO Rep *1*, 404-410.
Mayer, M.P. (2010). Gymnastics of molecular chaperones. Mol Cell *39*, 321-331.
Mayer, M.P., and Bukau, B. (2005). Hsp70 chaperones: cellular functions and molecular mechanism. Cell Mol Life Sci *62*, 670-684.
Milenkovic, D., Gabriel, K., Guiard, B., Schulze-Specking, A., Pfanner, N., and Chacinska, A. (2007). Biogenesis of the essential Tim9-Tim10 chaperone complex of mitochondria: site-specific recognition of cysteine residues by the intermembrane space receptor Mia40. J Biol Chem *282*, 22472-22480.
Misselwitz, B., Staeck, O., and Rapoport, T.A. (1998). J proteins catalytically activate Hsp70 molecules to trap a wide range of peptide sequences. Mol Cell *2*, 593-603.
Mogk, A., Schmidt, R., and Bukau, B. (2007). The N-end rule pathway for regulated proteolysis: prokaryotic and eukaryotic strategies. Trends Cell Biol *17*, 165-172.
Mokranjac, D., Berg, A., Adam, A., Neupert, W., and Hell, K. (2007). Association of the Tim14.Tim16 subcomplex with the TIM23 translocase is crucial for function of the mitochondrial protein import motor. J Biol Chem *282*, 18037-18045.
Mokranjac, D., and Neupert, W. (2008). Energetics of protein translocation into mitochondria. Biochim Biophys Acta.
Mokranjac, D., and Neupert, W. (2009). Thirty years of protein translocation into mitochondria: unexpectedly complex and still puzzling. Biochim Biophys Acta *1793*, 33-41.
Mokranjac, D., and Neupert, W. (2010). The many faces of the mitochondrial TIM23 complex. Biochim Biophys Acta *1797*, 1045-1054.

Moliere, N., and Turgay, K. (2009). Chaperone-protease systems in regulation and protein quality control in Bacillus subtilis. Res Microbiol *160*, 637-644.
Mollapour, M., Tsutsumi, S., and Neckers, L. (2010). Hsp90 phosphorylation, Wee1 and the cell cycle. Cell Cycle *9*, 2310-2316.
Morimoto, R.I. (2008). Proteotoxic stress and inducible chaperone networks in neurodegenerative disease and aging. Genes Dev *22*, 1427-1438.
Muchowski, P.J. (2002). Protein misfolding, amyloid formation, and neurodegeneration: a critical role for molecular chaperones? Neuron *35*, 9-12.
Muchowski, P.J., and Wacker, J.L. (2005). Modulation of neurodegeneration by molecular chaperones. Nat Rev Neurosci *6*, 11-22.
Munoz, I.G., Yebenes, H., Zhou, M., Mesa, P., Serna, M., Park, A.Y., Bragado-Nilsson, E., Beloso, A., de Carcer, G., Malumbres, M., Robinson, C.V., Valpuesta, J.M., and Montoya, G. (2011). Crystal structure of the open conformation of the mammalian chaperonin CCT in complex with tubulin. Nat Struct Mol Biol *18*, 14-19.
Murphy, M.P., Holmgren, A., Larsson, N.G., Halliwell, B., Chang, C.J., Kalyanaraman, B., Rhee, S.G., Thornalley, P.J., Partridge, L., Gems, D., Nystrom, T., Belousov, V., Schumacker, P.T., and Winterbourn, C.C. (2011). Unraveling the biological roles of reactive oxygen species. Cell Metab *13*, 361-366.
Murray, A.N., Solomon, J.P., Wang, Y.J., Balch, W.E., and Kelly, J.W. (2010). Discovery and characterization of a mammalian amyloid disaggregation activity. Protein Sci *19*, 836-846.
Nakai, T., Yasuhara, T., Fujiki, Y., and Ohashi, A. (1995). Multiple genes, including a member of the AAA family, are essential for degradation of unassembled subunit 2 of cytochrome c oxidase in yeast mitochondria. Mol. Cell. Biol. *15*, 4441-4452.
Nebauer, R., Schuiki, I., Kulterer, B., Trajanoski, Z., and Daum, G. (2007). The phosphatidylethanolamine level of yeast mitochondria is affected by the mitochondrial components Oxa1p and Yme1p. FEBS J *274*, 6180-6190.
Neupert, W., and Brunner, M. (2002). The protein import motor of mitochondria. Nat Rev Mol Cell Biol *3*, 555-565.
Neupert, W., and Herrmann, J.M. (2007). Translocation of proteins into mitochondria. Annu Rev Biochem *76*, 723-749.
Neutzner, A., Youle, R.J., and Karbowski, M. (2007). Outer mitochondrial membrane protein degradation by the proteasome. Novartis Found Symp *287*, 4-14; discussion 14-20.
Neuwald, A.F., Aravind, L., Spouge, J.L., and Koonin, E.V. (1999). AAA+: A class of chaperone-like ATPases associated with the assembly, operation, and disassembly of protein complexes. Genome Res *9*, 27-43.
Nijtmans, L.G., de Jong, L., Artal Sanz, M., Coates, P.J., Berden, J.A., Back, J.W., Muijsers, A.O., van der Spek, H., and Grivell, L.A. (2000). Prohibitins act as a membrane-bound chaperone for the stabilization of mitochondrial proteins. Embo J *19*, 2444-2451.
Nolden, M., Ehses, S., Koppen, M., Bernacchia, A., Rugarli, E.I., and Langer, T. (2005). The m-AAA protease defective in hereditary spastic paraplegia controls ribosome assembly in mitochondria. Cell *123*, 277-289.
Nollen, E.A., and Morimoto, R.I. (2002). Chaperoning signaling pathways: molecular chaperones as stress-sensing 'heat shock' proteins. J Cell Sci *115*, 2809-2816.

Nunnari, J., and Suomalainen, A. (2012). Mitochondria: in sickness and in health. Cell *148*, 1145-1159.
Ogura, T., Whiteheart, S.W., and Wilkinson, A.J. (2004). Conserved arginine residues implicated in ATP hydrolysis, nucleotide-sensing, and inter-subunit interactions in AAA and AAA+ ATPases. J Struct Biol *146*, 106-112.
Ogura, T., and Wilkinson, A.J. (2001). AAA+ superfamily ATPases: common structure--diverse function. Genes Cells *6*, 575-597.
Olzscha, H., Schermann, S.M., Woerner, A.C., Pinkert, S., Hecht, M.H., Tartaglia, G.G., Vendruscolo, M., Hayer-Hartl, M., Hartl, F.U., and Vabulas, R.M. (2011). Amyloid-like aggregates sequester numerous metastable proteins with essential cellular functions. Cell *144*, 67-78.
Ondrovicova, G., Liu, T., Singh, K., Tian, B., Li, H., Gakh, O., Perecko, D., Janata, J., Granot, Z., Orly, J., Kutejova, E., and Suzuki, C.K. (2005). Cleavage site selection within a folded substrate by the ATP-dependent lon protease. J Biol Chem *280*, 25103-25110.
Osman, C., Haag, M., Potting, C., Rodenfels, J., Dip, P.V., Wieland, F.T., Brugger, B., Westermann, B., and Langer, T. (2009a). The genetic interactome of prohibitins: coordinated control of cardiolipin and phosphatidylethanolamine by conserved regulators in mitochondria. J Cell Biol *184*, 583-596.
Osman, C., Merkwirth, C., and Langer, T. (2009b). Prohibitins and the functional compartmentalization of mitochondrial membranes. J Cell Sci *122*, 3823-3830.
Otera, H., and Mihara, K. (2011). Molecular mechanisms and physiologic functions of mitochondrial dynamics. J Biochem *149*, 241-251.
Palade, G.E. (1953). An electron microscope study of the mitochondrial structure. J Histochem Cytochem *1*, 188-211.
Paschen, S.A., Waizenegger, T., Stan, T., Preuss, M., Cyrklaff, M., Hell, K., Rapaport, D., and Neupert, W. (2003). Evolutionary conservation of biogenesis of beta-barrel membrane proteins. Nature *426*, 862-866.
Patel, S., and Latterich, M. (1998). The AAA team: related ATPases with diverse functions. Trends Cell Biol *8*, 65-71.
Pavlov, P.F., Wiehager, B., Sakai, J., Frykman, S., Behbahani, H., Winblad, B., and Ankarcrona, M. (2011). Mitochondrial gamma-secretase participates in the metabolism of mitochondria-associated amyloid precursor protein. FASEB J *25*, 78-88.
Pearce, D.A., and Sherman, F. (1995). Degradation of cytochrome oxidase subunits in mutants of yeast lacking cytochrome c and suppression of the degradation by mutation of yme1. J. Biol. Chem. *270*, 20879-20882.
Pearce, M.J., Mintseris, J., Ferreyra, J., Gygi, S.P., and Darwin, K.H. (2008). Ubiquitin-like protein involved in the proteasome pathway of Mycobacterium tuberculosis. Science *322*, 1104-1107.
Pfanner, N., Muller, H.K., Harmey, M.A., and Neupert, W. (1987). Mitochondrial protein import: involvement of the mature part of a cleavable precursor protein in the binding to receptor sites. Embo J *6*, 3449-3454.
Pickart, C.M., and Cohen, R.E. (2004). Proteasomes and their kin: proteases in the machine age. Nat Rev Mol Cell Biol *5*, 177-187.
Popov-Celcktic, D., Waegemann, K., Mapa, K., Neupert, W., and Mokranjac, D. (2011). Role of the import motor in insertion of transmembrane segments by the mitochondrial TIM23 complex. EMBO Rep *12*, 542-548.

Potting, C., Wilmes, C., Engmann, T., Osman, C., and Langer, T. (2010). Regulation of mitochondrial phospholipids by Ups1/PRELI-like proteins depends on proteolysis and Mdm35. EMBO J *29*, 2888-2898.

Powers, E.T., Morimoto, R.I., Dillin, A., Kelly, J.W., and Balch, W.E. (2009). Biological and chemical approaches to diseases of proteostasis deficiency. Annu Rev Biochem *78*, 959-991.

Pratje, E., and Guiard, B. (1986). One nuclear gene controls the removal of transient presequences from two yeast proteins: One encoded by the nuclear, the other by the mitochondrial genome. EMBO J. *5*, 1313-1317.

Rainey, R.N., Glavin, J.D., Chen, H.W., French, S.W., Teitell, M.A., and Koehler, C.M. (2006). A new function in translocation for the mitochondrial i-AAA protease Yme1: import of polynucleotide phosphorylase into the intermembrane space. Mol Cell Biol *26*, 8488-8497.

Rawlings, N.D., and Barrett, A.J. (1995). Evolutionary families of metallopeptidases. Methods Enzymol *248*, 183-228.

Rehling, P., Pfanner, N., and Meisinger, C. (2003). Insertion of hydrophobic membrane proteins into the inner mitochondrial membrane--a guided tour. J Mol Biol *326*, 639-657.

Reichert, A.S., and Neupert, W. (2004). Mitochondriomics or what makes us breathe. Trends Genet *20*, 555-562.

Retzlaff, M., Stahl, M., Eberl, H.C., Lagleder, S., Beck, J., Kessler, H., and Buchner, J. (2009). Hsp90 is regulated by a switch point in the C-terminal domain. EMBO Rep *10*, 1147-1153.

Richter, K., Haslbeck, M., and Buchner, J. (2010). The heat shock response: life on the verge of death. Mol Cell *40*, 253-266.

Roe, S.M., Ali, M.M., Meyer, P., Vaughan, C.K., Panaretou, B., Piper, P.W., Prodromou, C., and Pearl, L.H. (2004). The Mechanism of Hsp90 regulation by the protein kinase-specific cochaperone p50(cdc37). Cell *116*, 87-98.

Romisch, K. (2005). Endoplasmic reticulum-associated degradation. Annu Rev Cell Dev Biol *21*, 435-456.

Ross, C.A., and Poirier, M.A. (2004). Protein aggregation and neurodegenerative disease. Nat Med *10 Suppl*, S10-17.

Ross, C.A., and Poirier, M.A. (2005). Opinion: What is the role of protein aggregation in neurodegeneration? Nat Rev Mol Cell Biol *6*, 891-898.

Rottgers, K., Zufall, N., Guiard, B., and Voos, W. (2002). The ClpB homolog Hsp78 is required for the efficient degradation of proteins in the mitochondrial matrix. J Biol Chem.

Rowley, N., Prip-Buus, C., Westermann, B., Brown, C., Schwarz, E., Barrell, B., and Neupert, W. (1994). Mdj1p, a novel chaperone of the DnaJ family, is involved in mitochondrial biogenesis and protein folding. Cell *77*, 249-259.

Rugarli, E.I., and Langer, T. (2006). Translating m-AAA protease function in mitochondria to hereditary spastic paraplegia. Trends Mol Med *12*, 262-269.

Rugarli, E.I., and Langer, T. (2012). Mitochondrial quality control: a matter of life and death for neurons. EMBO J *31*, 1336-1349.

Rutherford, S.L., and Lindquist, S. (1998). Hsp90 as a capacitor for morphological evolution. Nature *396*, 336-342.

Sakoh, M., Ito, K., and Akiyama, Y. (2005). Proteolytic activity of HtpX, a membrane-bound and stress-controlled protease from Escherichia coli. J Biol Chem *280*, 33305-33310.

Sauer, R.T., and Baker, T.A. (2011). AAA+ proteases: ATP-fueled machines of protein destruction. Annu Rev Biochem *80*, 587-612.
Scheufler, C., Brinker, A., Bourenkov, G., Pegoraro, S., Moroder, L., Bartunik, H., Hartl, F.U., and Moarefi, I. (2000). Structure of TPR domain-peptide complexes: critical elements in the assembly of the Hsp70-Hsp90 multichaperone machine. Cell *101*, 199-210.
Schmidt, R., Bukau, B., and Mogk, A. (2009a). Principles of general and regulatory proteolysis by AAA+ proteases in Escherichia coli. Res Microbiol *160*, 629-636.
Schmidt, R., Zahn, R., Bukau, B., and Mogk, A. (2009b). ClpS is the recognition component for Escherichia coli substrates of the N-end rule degradation pathway. Mol Microbiol *72*, 506-517.
Schmitt, M., Neupert, W., and Langer, T. (1995). Hsp78, a Clp homologue within mitochondria, can substitute for chaperone functions of mt-hsp70. EMBO J. *14*, 3434-3444.
Schrader, E.K., Harstad, K.G., and Matouschek, A. (2009). Targeting proteins for degradation. Nat Chem Biol *5*, 815-822.
Shadel, G.S. (2005). Mitochondrial DNA, aconitase 'wraps' it up. Trends Biochem Sci *30*, 294-296.
Shah, Z.H., Hakkaart, G.A., Arku, B., de Jong, L., van der Spek, H., Grivell, L.A., and Jacobs, H.T. (2000). The human homologue of the yeast mitochondrial AAA metalloprotease Yme1p complements a yeast yme1 disruptant. FEBS Lett *478*, 267-270.
Shaw, J.M., and Nunnari, J. (2002). Mitochondrial dynamics and division in budding yeast. Trends Cell Biol *12*, 178-184.
Sherman, F. (1991). Getting started with yeast. Methods Enzymol *194*, 3-21.
Sherman, F., and Wakem, P. (1991). Mapping yeast genes. Methods Enzymol *194*, 38-57.
Sideris, D.P., and Tokatlidis, K. (2010). Oxidative protein folding in the mitochondrial intermembrane space. Antioxid Redox Signal *13*, 1189-1204.
Sikorski, R.S., and Hieter, P. (1989). A system of shuttle vectors and yeast host strains designed for efficient manipulation of DNA in Saccharomyces cerevisiae. Genetics *122*, 19-27.
Smith, M.H., Ploegh, H.L., and Weissman, J.S. (2011). Road to ruin: targeting proteins for degradation in the endoplasmic reticulum. Science *334*, 1086-1090.
Song, Z., Chen, H., Fiket, M., Alexander, C., and Chan, D.C. (2007). OPA1 processing controls mitochondrial fusion and is regulated by mRNA splicing, membrane potential, and Yme1L. J Cell Biol *178*, 749-755.
Stefani, M., and Dobson, C.M. (2003). Protein aggregation and aggregate toxicity: new insights into protein folding, misfolding diseases and biological evolution. J Mol Med (Berl) *81*, 678-699.
Steglich, G., Neupert, W., and Langer, T. (1999). Prohibitins regulate membrane protein degradation by the m-AAA protease in mitochondria. Mol. Cell. Biol. *19*, 3435-3442.
Steinmetz, L.M., Scharfe, C., Deutschbauer, A.M., Mokranjac, D., Herman, Z.S., Jones, T., Chu, A.M., Giaever, G., Prokisch, H., Oefner, P.J., and Davis, R.W. (2002). Systematic screen for human disease genes in yeast. Nat Genet *31*, 400-404.
Stiburek, L., Cesnekova, J., Kostkova, O., Fornuskova, D., Vinsova, K., Wenchich, L., Houstek, J., and Zeman, J. (2012). YME1L controls the accumulation of respiratory chain subunits and is required for apoptotic resistance, cristae morphogenesis and cell proliferation. Mol Biol Cell.

Stojanovski, D., Muller, J.M., Milenkovic, D., Guiard, B., Pfanner, N., and Chacinska, A. (2008). The MIA system for protein import into the mitochondrial intermembrane space. Biochim Biophys Acta *1783*, 610-617.
Strauss, K.M., Martins, L.M., Plun-Favreau, H., Marx, F.P., Kautzmann, S., Berg, D., Gasser, T., Wszolek, Z., Muller, T., Bornemann, A., Wolburg, H., Downward, J., Riess, O., Schulz, J.B., and Kruger, R. (2005). Loss of function mutations in the gene encoding Omi/HtrA2 in Parkinson's disease. Hum Mol Genet *14*, 2099-2111.
Striebel, F., Kress, W., and Weber-Ban, E. (2009). Controlled destruction: AAA+ ATPases in protein degradation from bacteria to eukaryotes. Curr Opin Struct Biol *19*, 209-217.
Studier, F.W., and Moffatt, B.A. (1986). Use of bacteriophage T7 RNA polymerase to direct selective high-level expression of cloned genes. J Mol Biol *189*, 113-130.
Suno, R., Niwa, H., Tsuchiya, D., Zhang, X., Yoshida, M., and Morikawa, K. (2006). Structure of the whole cytosolic region of ATP-dependent protease FtsH. Mol Cell *22*, 575-585.
Suzuki, C.K., Suda, K., Wang, N., and Schatz, G. (1994). Requirement for the yeast gene LON in intramitochondrial proteolysis and maintenance of respiration. Science *264*, 273-276.
Taanman, J.W., and Capaldi, R.A. (1992). Purification of yeast cytochrome c oxidase with a subunit composition resembling the mammalian enzyme. J Biol Chem *267*, 22481-22485.
Taipale, M., Jarosz, D.F., and Lindquist, S. (2010). HSP90 at the hub of protein homeostasis: emerging mechanistic insights. Nat Rev Mol Cell Biol *11*, 515-528.
Tamura, Y., Endo, T., Iijima, M., and Sesaki, H. (2009). Ups1p and Ups2p antagonistically regulate cardiolipin metabolism in mitochondria. J Cell Biol *185*, 1029-1045.
Tamura, Y., Onguka, O., Aiken Hobbs, A.E., Jensen, R.E., Iijima, M., Claypool, S.M., and Sesaki, H. (2012). Role for two conserved intermembrane space proteins, Ups1p and Up2p, in intra-mitochondrial phospholipid trafficking. J Biol Chem.
Tanaka, A., Cleland, M.M., Xu, S., Narendra, D.P., Suen, D.F., Karbowski, M., and Youle, R.J. (2010). Proteasome and p97 mediate mitophagy and degradation of mitofusins induced by Parkin. J Cell Biol *191*, 1367-1380.
Tatsuta, T. (2009). Protein quality control in mitochondria. J Biochem *146*, 455-461.
Tatsuta, T., Augustin, S., Nolden, M., Friedrichs, B., and Langer, T. (2007). m-AAA protease-driven membrane dislocation allows intramembrane cleavage by rhomboid in mitochondria. EMBO J *26*, 325-335.
Tatsuta, T., and Langer, T. (2007). Studying proteolysis within mitochondria. Methods Mol Biol *372*, 343-360.
Tatsuta, T., and Langer, T. (2008). Quality control of mitochondria: protection against neurodegeneration and ageing. EMBO J *27*, 306-314.
Tatsuta, T., and Langer, T. (2009). AAA proteases in mitochondria: diverse functions of membrane-bound proteolytic machines. Res Microbiol *160*, 711-717.
Taylor, J.P., Hardy, J., and Fischbeck, K.H. (2002). Toxic proteins in neurodegenerative disease. Science *296*, 1991-1995.
Thorsness, P.E., and Fox, T.D. (1993). Nuclear mutations in Saccharomyces cerevisiae that affect the escape of DNA from mitochondria to the nucleus. Genetics *134*, 21-28.
Thorsness, P.E., White, K.H., and Fox, T.D. (1993). Inactivation of YME1, a member of the ftsH-SEC18-PAS1-CDC48 family of putative ATPase-encoding

genes, causes increased escape of DNA from mitochondria in Saccharomyces cerevisiae. Mol Cell Biol *13*, 5418-5426.
Towbin, H., Staehelin, T., and Gordon, J. (1979). Electrophoretic transfer of proteins from polyacrylamide gels to nitrocellulose sheets: Procedure and some applications. Proc. Natl. Acad. Sci. USA *79*, 267-271.
Tyedmers, J., Mogk, A., and Bukau, B. (2010). Cellular strategies for controlling protein aggregation. Nat Rev Mol Cell Biol *11*, 777-788.
Valente, E.M., Salvi, S., Ialongo, T., Marongiu, R., Elia, A.E., Caputo, V., Romito, L., Albanese, A., Dallapiccola, B., and Bentivoglio, A.R. (2004). PINK1 mutations are associated with sporadic early-onset parkinsonism. Ann Neurol *56*, 336-341.
Vaughan, C.K., Gohlke, U., Sobott, F., Good, V.M., Ali, M.M., Prodromou, C., Robinson, C.V., Saibil, H.R., and Pearl, L.H. (2006). Structure of an Hsp90-Cdc37-Cdk4 complex. Mol Cell *23*, 697-707.
Vestweber, D., and Schatz, G. (1988). Point mutations destabilizing a precursor protein enhance its post-translational import into mitochondria. EMBO J. *7*, 1147-1151.
Voisine, C., Pedersen, J.S., and Morimoto, R.I. (2010). Chaperone networks: tipping the balance in protein folding diseases. Neurobiol Dis *40*, 12-20.
von der Malsburg, K., Muller, J.M., Bohnert, M., Oeljeklaus, S., Kwiatkowska, P., Becker, T., Loniewska-Lwowska, A., Wiese, S., Rao, S., Milenkovic, D., Hutu, D.P., Zerbes, R.M., Schulze-Specking, A., Meyer, H.E., Martinou, J.C., Rospert, S., Rehling, P., Meisinger, C., Veenhuis, M., Warscheid, B., van der Klei, I.J., Pfanner, N., Chacinska, A., and van der Laan, M. (2011). Dual role of mitofilin in mitochondrial membrane organization and protein biogenesis. Dev Cell *21*, 694-707.
von Janowsky, B., Knapp, K., Major, T., Krayl, M., Guiard, B., and Voos, W. (2005). Structural properties of substrate proteins determine their proteolysis by the mitochondrial AAA+ protease Pim1. Biol Chem *386*, 1307-1317.
Voos, W. (2012). Chaperone-protease networks in mitochondrial protein homeostasis. Biochim Biophys Acta.
Voos, W., and Rottgers, K. (2002). Molecular chaperones as essential mediators of mitochondrial biogenesis. Biochim Biophys Acta *1592*, 51.
Wach, A., Brachat, A., Alberti-Segui, C., Rebischung, C., and Philippsen, P. (1997). Heterologous HIS3 marker and GFP reporter modules for PCR-targeting in Saccharomyces cerevisiae. Yeast *13*, 1065-1075.
Walker, J.E., Eberle, A., Gay, N.J., Runswick, M.J., and Saraste, M. (1982a). Conservation of structure in proton-translocating ATPases of Escherichia coli and mitochondria. Biochem Soc Trans *10*, 203-206.
Walker, J.E., Saraste, M., Runswick, M.J., and Gay, N.J. (1982b). Distantly related sequences in the alpha- and beta-subunits of ATP synthase, myosin, kinases and other ATP-requiring enzymes and a common nucleotide binding fold. EMBO J *1*, 945-951.
Walter, P., and Ron, D. (2011). The unfolded protein response: from stress pathway to homeostatic regulation. Science *334*, 1081-1086.
Walter, S. (2002). Structure and function of the GroE chaperone. Cell Mol Life Sci *59*, 1589-1597.
Wandinger, S.K., Richter, K., and Buchner, J. (2008). The Hsp90 chaperone machinery. J Biol Chem *283*, 18473-18477.
Wang, K., and Klionsky, D.J. (2011). Mitochondria removal by autophagy. Autophagy *7*, 297-300.

Wasilewski, M., and Scorrano, L. (2009). The changing shape of mitochondrial apoptosis. Trends Endocrinol Metab *20*, 287-294.
Weber, E.R., Hanekamp, T., and Thorsness, P.E. (1996). Biochemical and functional analysis of the YME1 gene product, an ATP and zinc-dependent mitochondrial protease from S-cerevisiae. Mol. Biol. Cell *7*, 307-317.
Weber, E.R., Rooks, R.S., Shafer, K.S., Chase, J.W., and Thorsness, P.E. (1995). Mutations in the mitochondrial ATP synthase gamma subunit suppress a slow-growth phenotype of yme1 yeast lacking mitochondrial DNA. Genetics *140*, 435-442.
Wiedemann, N., Kozjak, V., Chacinska, A., Schonfisch, B., Rospert, S., Ryan, M.T., Pfanner, N., and Meisinger, C. (2003). Machinery for protein sorting and assembly in the mitochondrial outer membrane. Nature *424*, 565-571.
Xu, Z., Horwich, A.L., and Sigler, P.B. (1997). The crystal structure of the asymmetric GroEL-GroES-(ADP)7 chaperonin complex. Nature *388*, 741-750.
Yaguchi, T., Aida, S., Kaul, S.C., and Wadhwa, R. (2007). Involvement of mortalin in cellular senescence from the perspective of its mitochondrial import, chaperone, and oxidative stress management functions. Ann N Y Acad Sci *1100*, 306-311.
Yang, M., Jensen, R.E., Yaffe, M.P., Oppliger, W., and Schatz, G. (1988). Import of proteins into yeast mitochondria: the purified matrix processing protease contains two subunits which are encoded by the nuclear MAS1 and MAS2 genes. EMBO J. *7*, 3857-3862.
Yogev, O., Naamati, A., and Pines, O. (2011). Fumarase: a paradigm of dual targeting and dual localized functions. FEBS J *278*, 4230-4242.
Yogev, O., and Pines, O. (2011). Dual targeting of mitochondrial proteins: mechanism, regulation and function. Biochim Biophys Acta *1808*, 1012-1020.
Yoneda, T., Benedetti, C., Urano, F., Clark, S.G., Harding, H.P., and Ron, D. (2004). Compartment-specific perturbation of protein handling activates genes encoding mitochondrial chaperones. J Cell Sci *117*, 4055-4066.
Youle, R.J., and Narendra, D.P. (2011). Mechanisms of mitophagy. Nat Rev Mol Cell Biol *12*, 9-14.
Youle, R.J., and van der Bliek, A.M. (2012). Mitochondrial fission, fusion, and stress. Science *337*, 1062-1065.
Zhao, Q., Wang, J., Levichkin, I.V., Stasinopoulos, S., Ryan, M.T., and Hoogenraad, N.J. (2002). A mitochondrial specific stress response in mammalian cells. EMBO J *21*, 4411-4419.

7. ABBREVIATIONS

A	Ampère
AAA	triple A domain
AAC	ADP/ ATP carrier
ADP	adenosine diphosphate
Amp	ampicillin
APS	ammonium peroxodisulfate
ATP	adenosine triphosphate
ATPase	adenosine triphosphatase
BSA	bovine serum albumin
C-	carboxy-
CAT	carboxy atractyloside
CBB	coomassie brilliant blue
CCCP	carbonyl cyanide m-chlorophenylhydrazone
cDNA	complementary DNA
c-	centi-
Cyb2	Cytochrome b_2
DHFR	dihydrofolate reductase
DMSO	dimethylsulfoxid
DNA	desoxyribonucleic acid
DTT	dithiotreitol
Δ	delta
dNTP	deoxyribonucleoside triphosphate
ECL	enhanced chemiluminescence
E. coli	Escherichia coli
EDTA	ethylendiamine tetraacetate
g	gram
g	gravity
gDNA	genomic DNA

Hepes	N-2 hydroxyl piperazine-N'-2-ethane sulphonic acid
His6	hexahistidine polypeptide
Hsp	heat shock protein
H. sapiens	Homo sapiens
IgG	immunoglobulin G
IM	inner membrane
IMP	inner membrane peptidase
IMS	intermembrane space
IPTG	isopropyl-β,D-thiogalactopyranoside
kDa	kilo Dalton
kg	kilogram
l	liter
LB	Luria Bertani
LysC	endoprotease hydrolyzing at Lys-C-terminus
m	meter
m-	milli-
MBP	maltose-binding protein
min	minute
MPP	mitochondrial processing peptidase
MS	mass spectrometry
MTS	matrix targeting signal
MTX	methotrexate
M. musculus	Mus musculus
myc-	polypeptide derived from c-myc gene product
μ-	micro-
N-	amino-
NADH	nicotine amide adenine dinucleotide
NADPH	nicotine amide adenine dinucleotide phosphate
Ni-NTA	nickel-nitrilo triacetic acid
OD_x	optical density at x nm
OM	outer membrane
OXA	oxidase assembly protein

ABBREVIATIONS

PAGE	polyacrylamide gel electrophoresis
PCR	polymerase chain reaction
PEG	polyethylene glycol
PI	preimmune serum
PIR	protease inhibitor reagent (Roche)
PK	proteinase K
PMSF	phenylmethylsulfonylfluoride
RNA	ribonucleic acid
RNasin	ribonuclease inhibitor
rpm	rounds per minute
RT	room temperature
S. cerevisiae	Saccharomyces cerevisiae
SDS	sodium dodecyl sulfate
SILAC	stable isotope labeling of amino acids in cell culture
TBS	TRIS-buffered saline
TCA	trichloroacetic acid
TEMED	N.N.N',N'-tetramethylene diamine
TIM	translocase of the inner mitochondrial membrane
TOM	translocase of the outer mitochondrial membrane
TRIS	tris-(hydroxymethyl)-aminomethane
Tx-100	TritonX-100
U	Unit(s)
V	Volt
v/v	volume per volume
w/V	weight per volume
WT	wild-type
X_h	hydrophobic
$\Delta\psi$	membrane potential

1. ACKNOWLEDGEMENTS

Ich möchte mich ganz herzlich bei Professor Neupert bedanken. Es war eine großartige Möglichkeit für mich, in Ihrem Labor meine Doktorarbeit durchführen zu können. Vielen Dank für Ihren Glauben an mich und daran, dass ich es „noch besser" kann. So konnte ich mich weiterentwickeln, wachsen und wichtige Erfahrungen für die Zukunft sammeln.

Ich möchte mich bei Prof. Soll bedanken für seine Bereitschaft, mein offizieller Doktorvater zu sein. Vielen Dank, dass alle bürokratischen Schritte so unkompliziert verlaufen konnten.

Ich möchte mich bedanken bei Professor Vothknecht. Herzlichen Dank, dass Sie sich bereit erklärt haben, das Zweitgutachten meiner Doktorarbeit zu übernehmen und damit auch Mitglied meines Rigorosums zu sein. Ich möchte mich bei Frau Professor Conradt und Professor Felmy für ihr Interesse an meinem Projekt und die Teilnahme an meinem Rigorosum bedanken. Ich möchte mich bei Professor Nickelsen und Professor Schleicher für die unkomplizierte Durchführung des Umlaufs danken.

Many thanks to Professor Ladurner for giving me the opportunity to continue working in the institute that got so nicely intercultural and interdisciplinary creating a highly motivating working atmosphere.

In particular, I wish to thank Dejana for being my patient and motivating group leader through "good and bad" times, always encouraging me in the right moment, teaching me that patience is gold in science and that being deliberate is more important than being fast. Thank you for sharing your enormous scientific knowledge with me and furthermore for taking care of my non-scientific gaps of knowledge (Who is Darth Vader?).

Ich möchte mich beim Elitenetzwerk Bayern bedanken für die Möglichkeit am Graduiertenkolleg 'Protein Dynamics in Health and Disease' teilnehmen zu können. Das Programm war für meine wissenschaftliche Entwicklung von großer Bedeutung.

Vielen Dank, liebe Christine, für deine stetigen Bemühungen und deine Unterstützung in allen bürokratischen Angelegenheiten. Ich war immer froh, eine Anlaufstelle für all meine Fragen und Probleme zu haben. Danke dir, Kai, für deine Unterstützung in wissenschaftlichen Fragen, für dein großes Engagement beim ENB und viele Runden im Park. Thank you Corey for your time and patience in helping me to make my rather poetic and literary writing a bit more scientific. Hope, you didn't suffer too much! Thank you, Francois for your endless patience in helping me with all kinds of computer and any other problems, your advice in professional issues and lots of fun in the lab. Karin und Petra, danke euch Mädels für viele schöne, lustige Stunden im Labor und jede Menge Unterstützung und Motivation wenn es mal wieder nicht so lief. Liebe Chrissy, wir haben uns erst spät besser kennengelernt und trotzdem hab ich das Gefühl, dass wir uns schon ewig kennen. Danke für so viele gute Gespräche, nachdenkliche und lustige Stunden, für deine Gesellschaft bei Zumba, Rock im Park und sonstigen Aktionen. Ich hoffe sehr, dass wir in Kontakt bleiben. Liebe Simone, vielen Dank für die schöne Zeit hier in München. Wir hatten so viele gute Gespräche über „Gott und die Welt". Ich wünsch mir, dass wir uns nicht aus den Augen verlieren trotz großer Entfernung.

Vielen Dank, liebe Christiane für deine Freundschaft, die ein so wichtiger Bestandteil für meine "Münchner Zeit" war, sowohl beruflich als auch privat. Danke für viele gute Gespräche im Auto bei laufendem Motor, bei unserem Da Ugo oder

ACKNOWLEDGEMENTS

beim joggen... ☺ Ich werd's sehr vermissen und hoffe, wir bleiben in Kontakt. Danke dir Peter für nette „Werkstattgespräche" und immer aufbauende Worte, für deine Hilfsbereitschaft und für unzählige Fahrrad-Notoperationen.

Liebe Heike, dir auch ein ganz herzliches Dankeschön. Mit viel Zeit und Geduld hast du mich ins Projekt eingearbeitet. Und es motiviert so sehr, wenn man auch für "kleine Dinge", z. B. eine funktionierende Klonierung, gelobt wird! Liebe Nikola, auch dir lieben Dank für deine Unterstützung im Labor und die Geduld mit der du während all der Jahre meine unzähligen Fragen beantwortet hast. Und natürlich vielen Dank für die tolle Freundschaft mit massenhaft (entkoffeinierten!) MILCH-Kaffees und Shopping-Touren. Ich wünsch mir sehr, dass wir in Kontakt bleiben können. Mittlerweile hast du dann zwei Anlaufstellen in Stockholm. ☺ Liebe Regina, vielen Dank für die vielen schönen Stunden, die wir zusammen auf dem Berg, im Wasser, bei Kaffee und Kuchen oder auf dem Spielplatz verbracht haben. Ich hab mich bei euch immer so willkommen und zu Hause gefühlt. Ich hoffe sehr, dass wir uns aber trotz der Entfernung nie aus den Augen verlieren. Lieber Max, danke dir auch für deine Unterstützung in sämtlichen wissenschaftlichen Fragen, für jede Menge Antikörper und für viel Spaß im Labor.

I would like to thank all former and present members of the Neupert and Ladurner lab. I really want to thank EACH one of you as everyone contributes or contributed to the great atmosphere in the lab. I loved coming to the lab from the first day till the last day and that is not only due to the project! It feels like meeting friends everyday and makes hard times with the project so much easier.

Und meine lieben Freunde in der Ferne, Verena, Julia und Cristina, Eva, Gretel und Rese, Nadine, Lukas - auch euch ein großes Dankeschön. Ich freu mich so, dass der Kontakt geblieben ist, dass wir uns trotz teils größerer Entfernung nicht "auseinander gelebt" haben, uns immer wieder sprechen und es so ist „wie immer", wenn wir uns wieder treffen.

Nicht zuletzt möchte ich meiner Familie danken. Ihr habt mich immer bestärkt und an mich geglaubt, auch nach allem, was vorher war. Ich hab eine zweite Chance bekommen und ihr habt mich darin unterstützt, sie zu nutzen. Ich wünsch mir sehr, dass wir auch in Zukunft weiter so viel Spaß miteinander haben und uns so nah sind, auch wenn wir uns nicht oft sehen können.

i want morebooks!

Buy your books fast and straightforward online - at one of world's fastest growing online book stores! Environmentally sound due to Print-on-Demand technologies.

Buy your books online at
www.get-morebooks.com

Kaufen Sie Ihre Bücher schnell und unkompliziert online – auf einer der am schnellsten wachsenden Buchhandelsplattformen weltweit! Dank Print-On-Demand umwelt- und ressourcenschonend produziert.

Bücher schneller online kaufen
www.morebooks.de

VDM Verlagsservicegesellschaft mbH
Heinrich-Böcking-Str. 6-8
D - 66121 Saarbrücken

Telefon: +49 681 3720 174
Telefax: +49 681 3720 1749

info@vdm-vsg.de
www.vdm-vsg.de

Printed by Books on Demand GmbH, Norderstedt / Germany